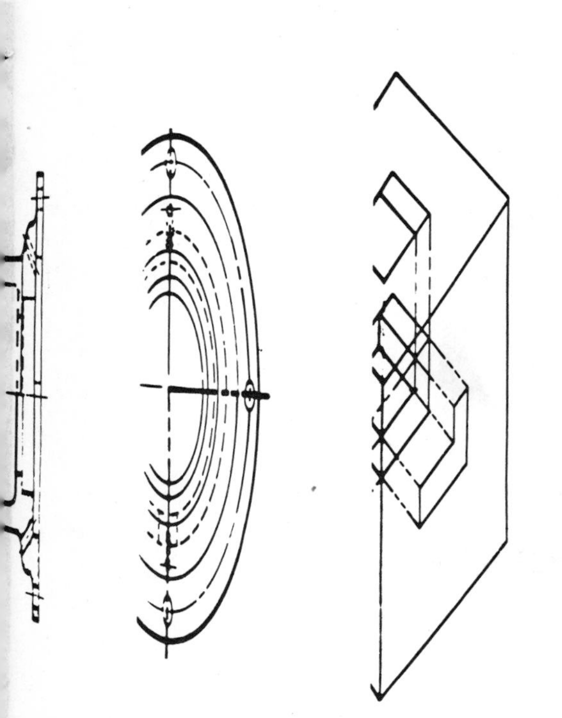

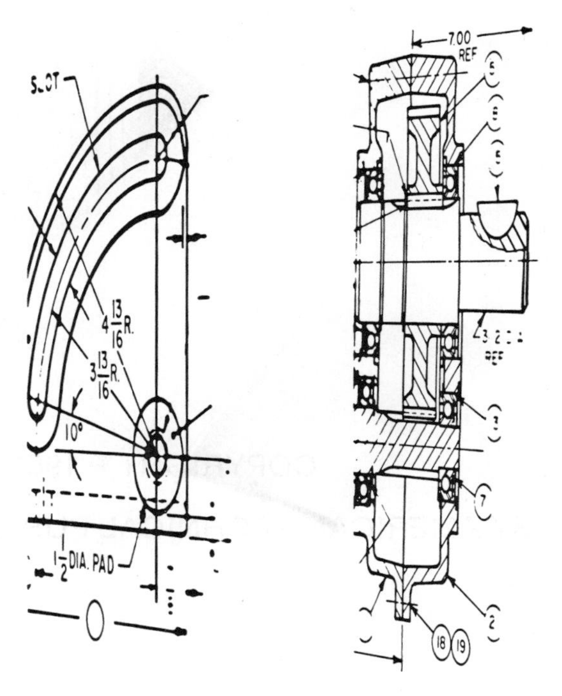

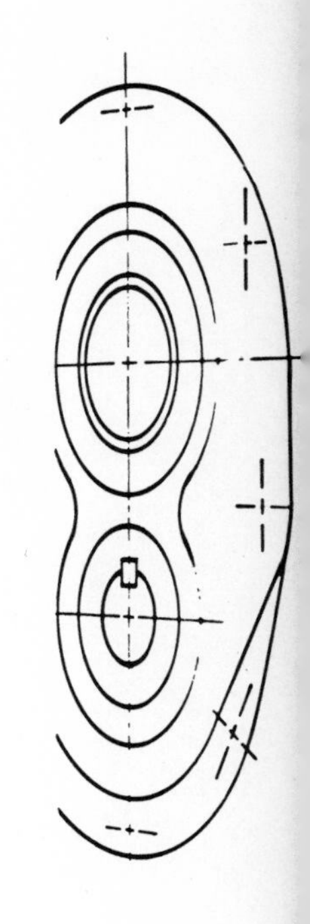

print reading for machine shop

ronald d. riggs

greenville technical college

american
technical
publishers, inc. alsip, illinois 60658

Library of Congress Catalog Number: 81-67536
ISBN: 0-8269-1870-0

123456789-81-987654321

PRINTED IN THE UNITED STATES OF AMERICA

PREFACE

PRINT READING FOR MACHINE SHOP is designed to give a basic introduction to reading machine trade blueprints. No prior knowledge of print reading is assumed or needed. The text starts with basic principles and progresses to more complex industrial prints. Interpretation of prints is stressed throughout. The text is heavily illustrated to show *how* to interpret the print.

Twenty-six Self Check Quizzes allow students to check their own progress. Answers to the Self Check Quizzes are given at the end of the book. Fifty-three print reading exercises and three Trade Competency Tests are included.

Sketching is introduced early in the text. Sketching allows the student to become familiar with basic drafting conventions and practices.

A math review covers the basic math used for interpreting a blueprint. Measurement is also covered.

Metric conventions and prints are introduced in a special unit.

PRINT READING FOR MACHINE SHOP covers the key conventions, symbols, and abbreviations found in today's industry. The numerous quizzes and exercises help the student to achieve basic blueprint reading competency.

The author and the publisher wish to express special appreciation to Marshall & Williams Company and James C. White Company for allowing use of their company drawings.

THE PUBLISHERS

contents

unit		page
1.	prints: industry's means of communication	1
2.	content of a print: information for the worker	3
3.	the alphabet of lines	13
4.	orthographic projection	21
5.	freehand sketching	33
6.	pictorial drawings	43
7.	drawing scales	53
8.	math review	59
9.	dimensions	71
10.	dimensioning holes	79
11.	limits and tolerances	89
12.	section views	93
13.	auxiliary views	103
14.	threads and fasteners	109
15.	surface finish	117
16.	conventional drafting practices	123
17.	metrics	129
18.	additional exercises	139
	trade competency tests	151
	answers to self-check quizzes	157
	glossary	162
	index	164

prints: industry's means of communication

Unit 1

After completing this unit you will be able to explain how a drawing is made and why prints of the drawing are made.

what is a print?

We communicate in many different ways. We usually discuss news, sports, and other items of interest face-to-face. Sometimes we communicate by telephone. Other times we write a letter. Song writers communicate by lyrics and music. In industry, however, the most widely used means of communication is the *print* (Figure 1-1).

What is a print? A print is a copy of a drawing. It conveys information. It communicates. A print is a message of building information from designer to manufacturer. You can expect to find many different kinds of information on a print. Figure 1-2 shows a print with five different "messages" for the worker.

figure 1-1
on-the-job communications

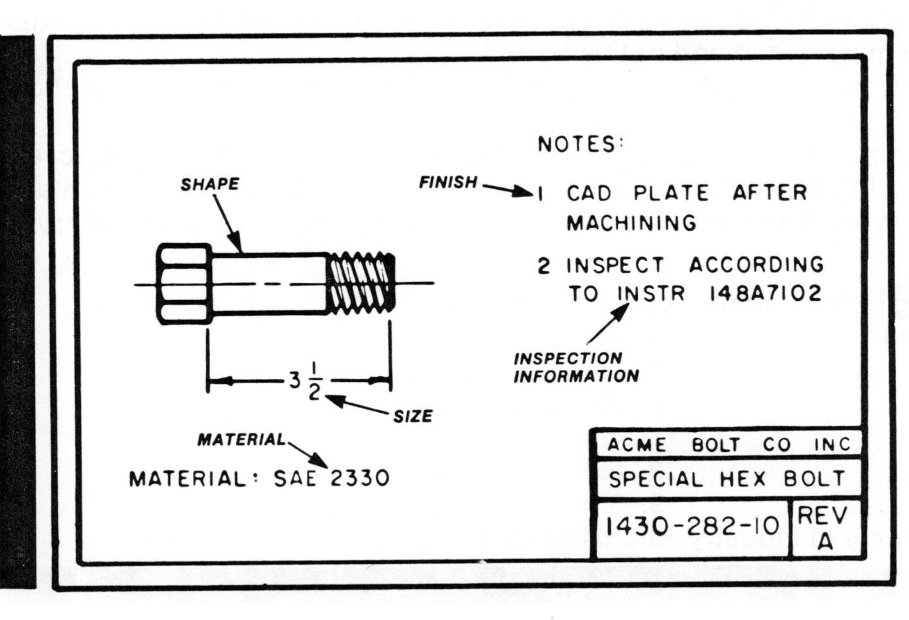

figure 1-2
typical "messages" communicated by a print

We use the terms *blueprint, print,* and *drawing* interchangeably. The term *blueprint* originally meant a *reproduction* (copy) with white lines and blue background. Today, the term *blueprint* is used for prints made by any of several different reproduction processes. Usually the print has blue, black, or brown lines on a white background. In the machine trades, we use the term *prints* instead of *blueprints.*

1. *A NEED ARISES*
2. *AN IDEA IS CONCEIVED*
3. *SKETCHES AND LAYOUTS ARE MADE*
4. *A DRAWING IS CREATED*
5. *SHOP DETAIL DRAWINGS ARE PRODUCED*
6. *PRINTS ARE MADE FOR THE SHOP*

figure 1-3
the evolution of a print

the evolution of a print

How does a print come about? First, a need arises. That need is communicated to someone in the design department such as a designer. The designer comes up with an idea to meet the need. Then the designer makes sketches and layouts. From the sketches, a formal drawing is created. Finally, detail drawings or shop drawings are made. From these, prints are made and given to the worker. See Figure 1-3. What begins as a need expressed in word form results in a drawing containing the technical information necessary to satisfy the need. Prints are the language of industry. Engineers, architects, designers, drafters, technical writers, and workers in the various trades all communicate by prints. In order to succeed in industry, you must be able to communicate in this language.

self-check quiz 1-a

Choose the *best* answer without looking back. Then review the unit to check your answers. (Correct answers are given at the end of this book.)

1. Telegrams, prints, music, and newspapers are all examples of (a) education; (b) communication; (c) investigation; (4) installation.
2. The main purpose of a print is to (a) explain an idea; (b) give a pleasing picture; (c) convey a message of building or machining information; (d) document what has been done.
3. A print may contain (a) shape description; (b) size information; (c) assembly instructions; (d) all of these.

Arrange the following in the correct order of evolution:

A. Formal Drawing	4. 1st ________
B. Idea	5. 2nd ________
C. Need	6. 3th ________
D. Sketches	7. 4th ________
E. Detail Drawings	8. 5th ________
F. Prints	9. 6th ________

10. List 6 types of people in industry who use prints.

content of a print: information for the worker

Unit 2

After completing this unit, you will be able to identify and explain the information found on a print.

In order to machine an object, you need more than a picture of it on a drawing. The drawing must have a number and a name. Views of the object must show dimensions. Different parts on the drawing must have part numbers. Finally, other information in the form of notes, references, and revision status completes the drawing. Each of these items on a drawing provides vital information for the worker.

drawing number

The *drawing number* identifies the drawing. Machine shop drawings may be identified by numbers only (Figure 2-1A) or by numbers and letters (Figure 2-1B and C). Sometimes the letter in a drawing number indicates the size of the drawing. In Figure 2-1B, the *A* in the drawing number indicates an "A" size drawing.

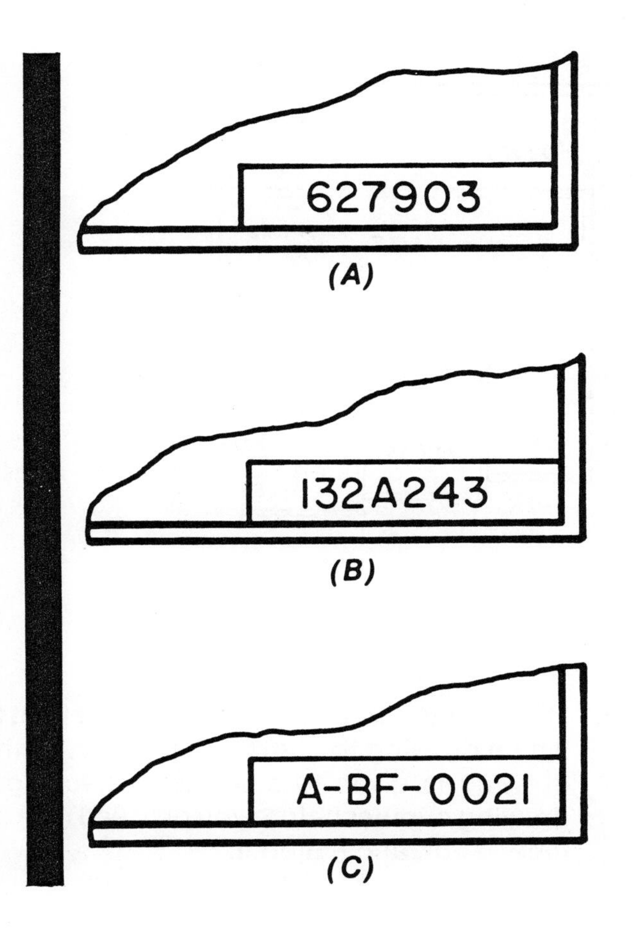

figure 2-1
typical drawing numbers

CUSTOMARY SHEET SIZE	DIMENSIONS INCHES (MILLIMETERS)	COMPARABLE METRIC SIZE	DIMENSIONS MILLIMETERS
A	8.5x11 (216x280) or 9x12 (229x305)	A4	210x297
B	11x17 (280x432) or 12x18 (305x457)	A3	297x420
C	17x22 (432x559) or 18x24 (457x610)	A2	420x594
D	22x34 (559x864) or 24x36 (610x914)	A1	594x841
E	34x44 (864x1118) or 36x48 (914x1219)	A0	841x1189

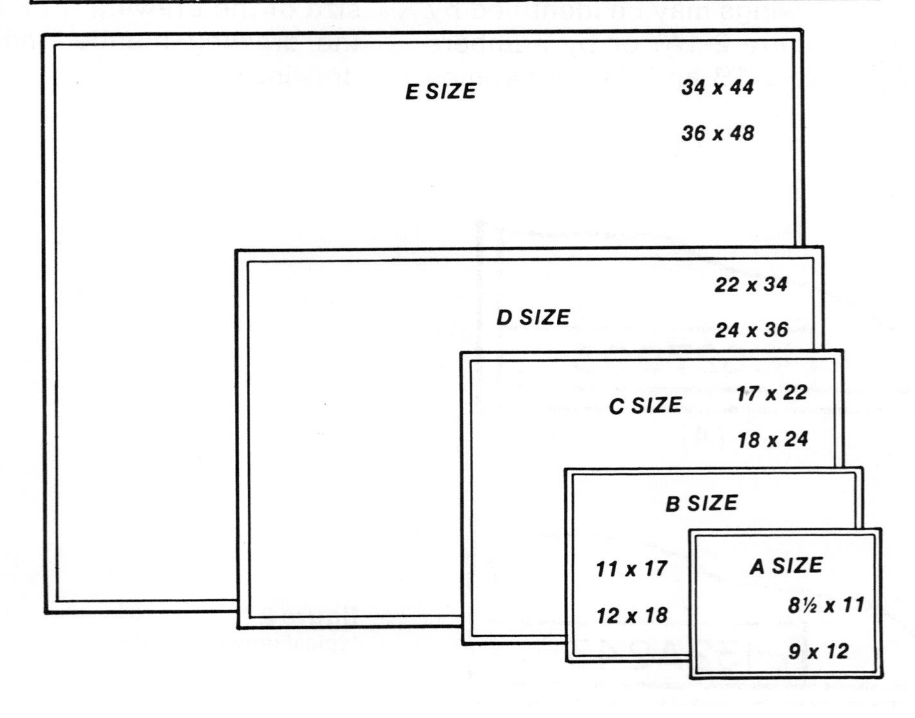

figure 2-2
a comparison of standard drawing sheet sizes

Figure 2-2 shows the standard drawing sizes used in industry.

The first step in interpreting or "reading" a print is confirming the drawing number. Most trades use planning sheets which tell them step-by-step what to do. A planning sheet tells which drawing to use. Occasionally, a wrong drawing is given to a worker. Checking the drawing number before you proceed may save time, labor, and material.

drawing title

The second step in reading a print is to examine the title. The *drawing title* tells the name of the object shown on the drawing. Sometimes the title indicates the purpose of the drawing. For example, in the title shown in Figure 2-3, the object shown on the drawing is an exhaust duct. The purpose of the drawing is to tell how to fabricate (build) and machine the exhaust duct.

EXHAUST DUCT
FABRICATION & MACHINING

figure 2-3
a typical drawing title

(A) INFORMAL	***(B) FORMAL***
COVER PLATE	***PLATE, COVER***
ENGINE SKID MOUNTING PAD	***PAD, MOUNTING — ENGINE SKID***
INPUT GEAR SUPPORT ASSEMBLY	***SUPPORT, ASSEMBLY OF-INPUT GEAR***

figure 2-4
different ways of wording titles

A drawing title may be informal as shown in Figure 2-4A, or it may be formal as in Figure 2-4B. Some industries have precise instructions on how titles must be worded. Other industries (like the construction industry) have developed an informal wording format which everyone generally follows. The title is usually placed in the lower right-hand corner of the drawing in the title block along with other important information (Figure 2-5).

The drawing title gives you clues about what you will find on the print. Reading the title prepares you for the job of printreading. Reading the title seems like a simple thing to do, and it is! However, a careless worker often skips this very important step of printreading. Discipline yourself to seek out the title and analyze it whenever you pick up a print.

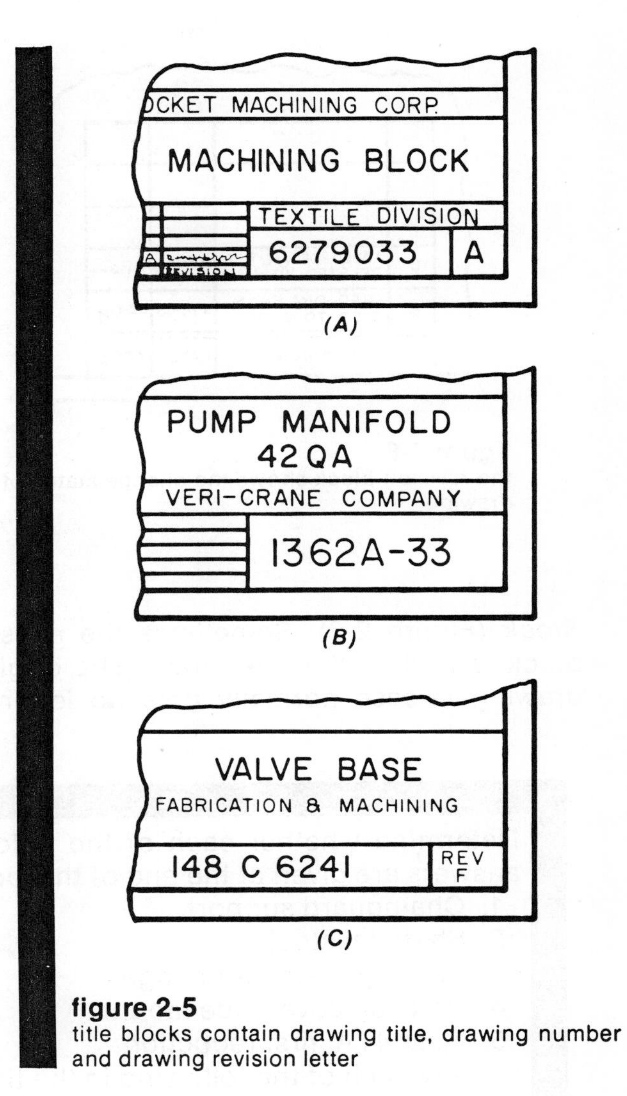

figure 2-5
title blocks contain drawing title, drawing number and drawing revision letter

drawing revisions

It is often necessary to make drawing changes or *revisions* after prints have been issued. There are many reasons for this. Sometimes drawings contain errors. Sometimes parts shown on drawings become obsolete and need to be replaced by newer versions. Someone may figure out a better way to make the part. In all these cases, the facts must be brought to the attention of the engineer, designer, or drafter so the drawing can be revised.

The revision status is usually shown by a letter or number in the *Revision Block* or *Change*

B	ADDED DETAIL F-2. DELETED VIEW A-A	5-27-79	R. Putman
A	CHGD. DIM. FROM 2.50 TO 2.48	3-14-78	R.D.Riggs
LTR.	REVISIONS	DATE	APP'D.

figure 2-6
the revision block shows the change status of the drawing

Block (Figure 2-6). Sometimes the revision block is part of the title block. The original drawing release normally gets no letter or number designation, and the revision block is left blank or filled in with a dash (—). The first revision of the drawing is called Revision A. Subsequent revisions become Revision B, Revision C, and so on. Some companies use a numbering system such as Revision 1, Revision 2, and so on.

The worker must always be sure the latest revision is being used in the shop. Much time, effort, and money are wasted when a worker builds or machines to an obsolete (out-of-date) print. When referring to a print it is a good practice to always mention the revision you are working to. Say, "On drawing 148A7310, Revision C, it shows that...."

On the views of the revised drawing, a triangle or circle is placed near the detail, dimension or note that was changed. The revision letter or number is placed inside the triangle or circle.

self-check quiz 2-a

Determine whether each of the following drawing titles is formal or informal. (Correct answers are given at the end of this book.)

1. Chainguard support ________
2. Shaft, tailstock ________
3. Turbine casing aft flange ________
4. Bracket, cover-side mount ________
5. Gear 4TA, first reduction ________

Identify each of the following in the title block shown.

6. Drawing number ________
7. Drawing title ________
8. Drawing revision ________
9. Drawing size ________
10. What was changed when the drawing was revised? ________

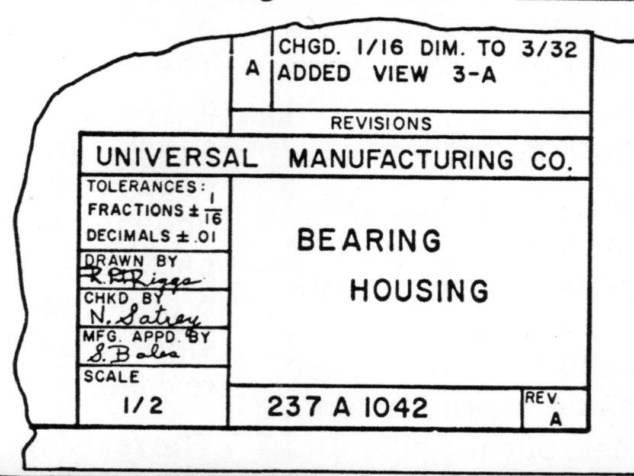

views

Views show the size and shape of the objects on the drawing. The drafter shows only the views necessary to fully describe the object. All unnecessary views are omitted. Sometimes a single view will do the job. Usually it takes two or three views. For complicated parts, auxiliary views and internal (section) views are needed. Normally a front view of an object is the main or principal view. You will learn more about these views later.

Analyzing the views on a print is the hardest part of printreading. Sometimes, understanding the print is harder than making the part! Flat, two-dimensional views must be studied until a three-dimensional object is pictured in your mind. Features on the views, such as holes and protrusions, will help you to visualize the three-dimensional object.

notes

Views of an object do not always give all the information you need. Additional information is given on drawings in the form of *notes*. Often, important, explanatory material is more easily expressed in words. Typical notes are shown in Figure 2-7.

Notes should be skimmed before the views of the drawing are analyzed. The notes will forewarn you of special requirements not readily seen in the views. Important information in the notes is often missed by the worker. This results in incomplete or incorrect work. Skim the notes first. They will help you analyze the views correctly.

part numbers

When a drawing contains more than one item or part, each part is assigned a number. Usually, the most important (or largest) part is given part number 1. The next most important (or next larger) part is assigned part number 2, and so on. Part numbers are defined in the *Bill of Material* (Figure 2-8) on the drawing. The Bill of Material is sometimes called a *Parts List*, *List of Material*, or *Equipment Schedule*.

DEBURR AND BREAK ALL CORNERS
MAKE FROM 1.50 BAR STOCK
ALL MACHINED RADII TO BE .14-.10 UNLESS OTHERWISE SPECIFIED
ALL CHAMFERS TO BE .14-.10 x 46-44° UNLESS OTHERWISE SPECIFIED
FASTEN ROD INTO MOUNTING ARM AND MATCHDRILL .25 DIA HOLE

figure 2-7
typical notes found on machine shop drawings

PART NO.	REQD	DESCRIPTION	MATL SPEC	REMARKS
3	2	BAR, 5/8 SQ x 3.00 LG	CRS	KEY
2	1	PLATE, .125 x 14.00 DIA	316SS	
1	1	TUBING, 14.00 DIA x .50 WALL	316SS	
BILL OF MATERIAL				

figure 2-8
a typical bill of material.

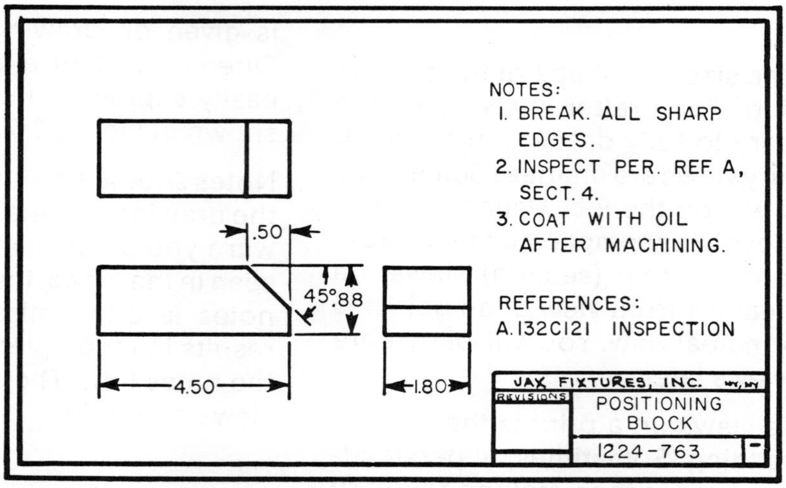

NOTES:
1. BREAK. ALL SHARP EDGES.
2. INSPECT PER. REF. A, SECT. 4.
3. COAT WITH OIL AFTER MACHINING.
REFERENCES:
A. 132C121 INSPECTION
.50
45°
.88
4.50
1.80
JAX FIXTURES, INC.
REVISIONS
POSITIONING BLOCK
1224-763

figure 2-9
a drawing showing notes and reference for the worker.

4
3
NOTES
1. BREAK EDGES .060
2. MAKE FROM B-7212-12
3 COAT WITH ENSIS 20 AFTER FAB.
.50
1.3
1.3
A
B-774-2-12
TITLE
SHIM FABRICATION
REVISIONS
DRAWING NO.
B-774-2-12
REV
A
5
1
2

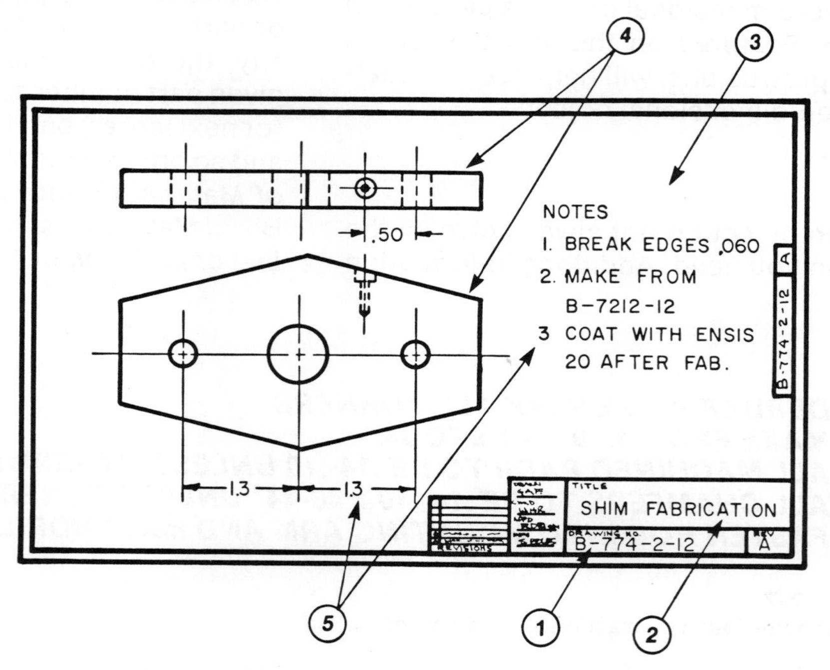

1. CONFIRM DRAWING NUMBER AND REVISION.
2. ANALYZE TITLE.
3. SKIM NOTES.
4. ANALYZE VIEWS.
A. MENTALLY IDENTIFY EACH VIEW (FRONT, SIDE, TOP AND SO ON).
B. STUDY THE FRONT VIEW.
C. IDENTIFY SURFACES OF THE FRONT VIEW IN OTHER VIEWS.
D. EXAMINE SPECIAL FEATURES SUCH AS CURVED PORTIONS, HOLES AND THREADS. LOCATE THESE FEATURES IN RELATED VIEWS.
5. AFTER THE OBJECT BEGINS TO TAKE SHAPE IN YOUR MIND, ANALYZE SUPPLEMENTARY INFORMATION SUCH AS DIMENSIONS, NOTES AND REFERENCES.

figure 2-10
procedure for analyzing a drawing.

references

A drawing may refer to other drawings or documents which contain important information for the worker. These other documents are called *references*. A machining drawing may reference heat treating specifications or inspection requirements (Figure 2-9). A welding drawing may reference a welding specification which defines the welding process and procedures to use. The requirements of the references are part of the drawing just as if they were shown on the drawing itself.

dimensions

Views on a drawing permit you to visualize the part. However, to make the part, you need dimensions. Dimensions are discussed in detail in Unit 9. Figure 2-9 illustrates some simple decimal dimensions. Unless otherwise specified, dimensions are in *inches*. The 4.50 dimension in Figure 2-9 means 4.50" or 4½ inches.

summary

Figure 2-10 summarizes the proper procedure for analyzing a drawing. You should recognize each of the drawing components in relation to the summary given below:

A. DRAWING NUMBER—Identifies the drawing.
B. DRAWING TITLE—Names the object shown on the drawing.
C. TITLE BLOCK—Contains the title, drawing number, and other related information.
D. REVISION BLOCK—Shows the revision status of the drawing.
E. NOTES—Give additional information.
F. VIEWS—Describe fully the shape of the object.
G. PART NUMBERS—Identify individual pieces which make up the object. Parts are identified in a PARTS LIST or BILL OF MATERIAL.
H. REFERENCES—Identify other documents which contain relevant information.
I. DIMENSIONS—Define the size of the object.
J. BILL OF MATERIAL—Shows a table of parts with identifying information.

self-check quiz 2-b

Circle T if the statement is true or F if it is false. (Answers are given at the end of this book.)

1. Analyzing the drawing title requires self-discipline. T F
2. Notes should be skimmed before analyzing views. T F
3. It is not important to identify each view of a drawing. T F
4. The front view is usually the principal view. T F
5. Features such as holes, protrusions and threaded portions help to relate one view to another. T F
6. The objective of printreading is to form a two-dimensional picture in your mind. T F
7. Analyzing dimensions is one of the last steps in printreading. T F

This page for student notes

exercise 2-1

____________________ name

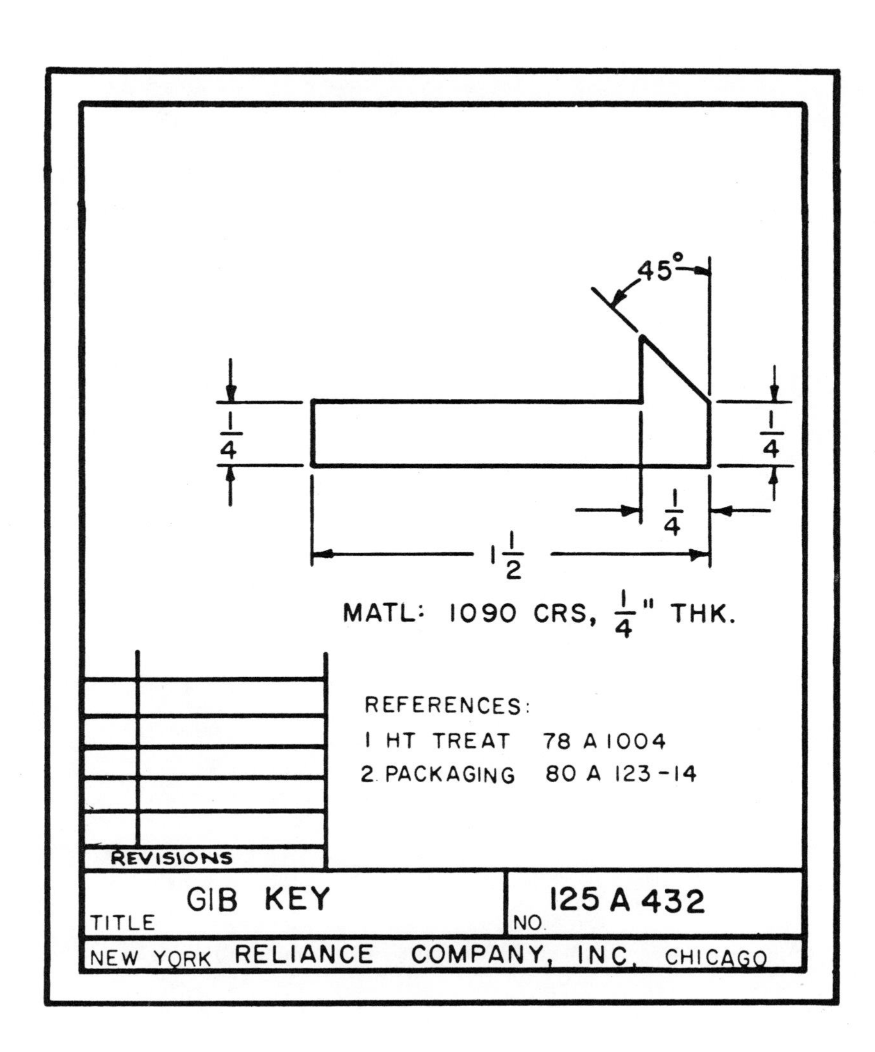

1. What is the drawing number? ________
2. What is the title? ________
3. What material is to be used? ________
4. How long is the part? ________
5. How thick (deep) is the part? ________
6. What specification is to be used to package the part after machining? ________
7. What document calls out the heat treating requirements? ________
8. Has the drawing been revised? ________
9. Is the title formal or informal? ________
10. What size is the drawing? ________

This page for student notes

the alphabet of lines

Unit 3

After completing this unit, you will be able to recognize different types of lines used on drawings.

A drafter uses different lines to help visualize objects on a drawing. The American National Standards Institute (ANSI) recommends a system of lines of different thicknesses, lengths and appearances. This system is called the *alphabet of lines.* It includes object lines, hidden lines, center lines, extension lines, dimension lines and other special purpose lines.

object lines

The basic shape of an object is defined by solid, heavy lines known as *visible object lines,* or simply *object lines*. Object lines (Figure 3-1) outline the object and show any distinct change in shape.

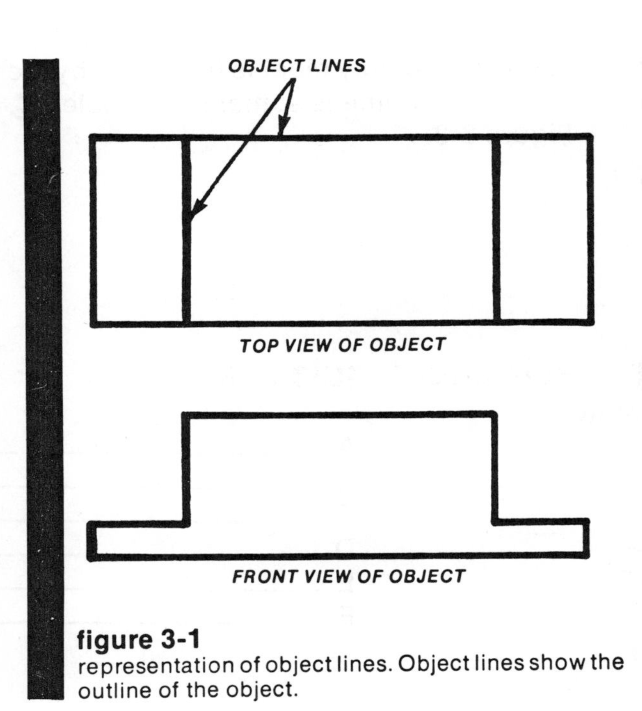

figure 3-1
representation of object lines. Object lines show the outline of the object.

hidden lines

To fully visualize an object, the worker needs to know where all surfaces and edges of the object are located. Short, dashed lines (------) represent features hidden from view. These lines are called *hidden lines* (Figure 3-2). Hidden lines show edges hidden behind a surface in a view.

HIDDEN LINES

TOP VIEW OF OBJECT

FRONT VIEW OF OBJECT

figure 3-2
representation of hidden lines. You can't see the inside corners when looking at the top view. Hidden lines show an edge hidden in a view.

center lines

Lines passing through the center of an object are known as *center lines*. Center lines consist of alternating long and short, thin lines

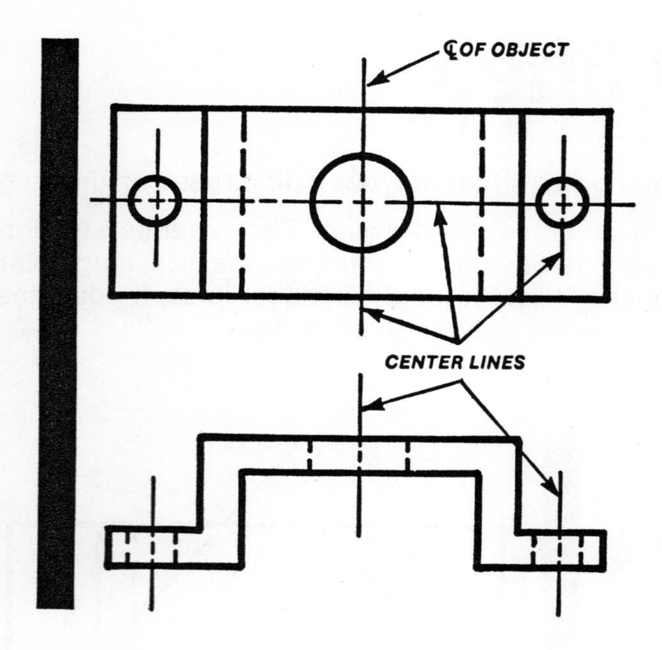

figure 3-3
representation of center lines

(———————— — ————————— — ————). Center lines also are used to indicate the center of a circle, hole, or curved portion of an object.

When two center lines intersect (cross), the place where they meet is indicated by a cross (+). A center line is sometimes labeled ℄. See Figure 3-3.

self-check quiz 3-a

Identify each of the lines on the figure below:

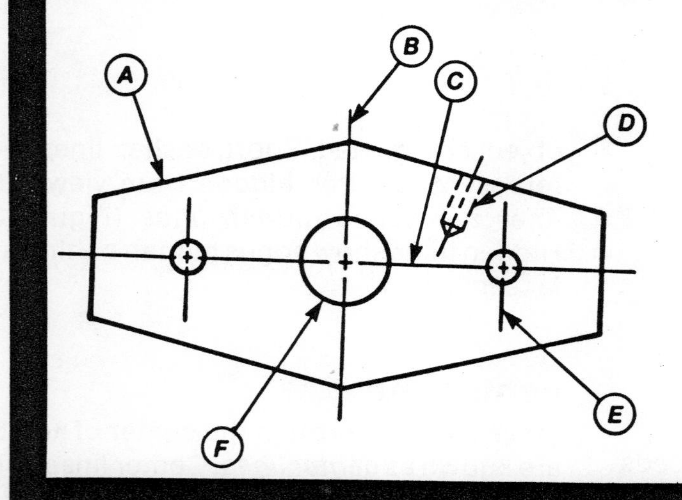

A ____________________
B ____________________
C ____________________
D ____________________
E ____________________
F ____________________

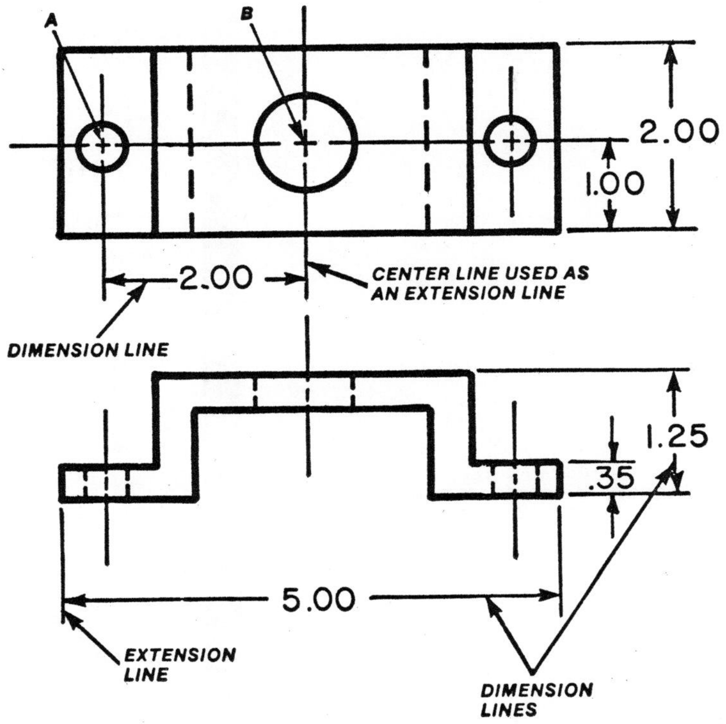

figure 3-4
representation of extension and dimension lines

extension and dimension lines

Dimensions are needed to give the size of an object. Thin, solid lines called *extension lines* are placed on a drawing to define the portion of the object being measured. Dimension lines are perpendicular to the extension lines (Figure 3-4). *Dimension lines* are also thin, solid lines. They are parallel to the dimension they define. Normally the dimension is placed midway along the dimension line. In Figure 3-4 the distance from point A to point B is 2 inches.

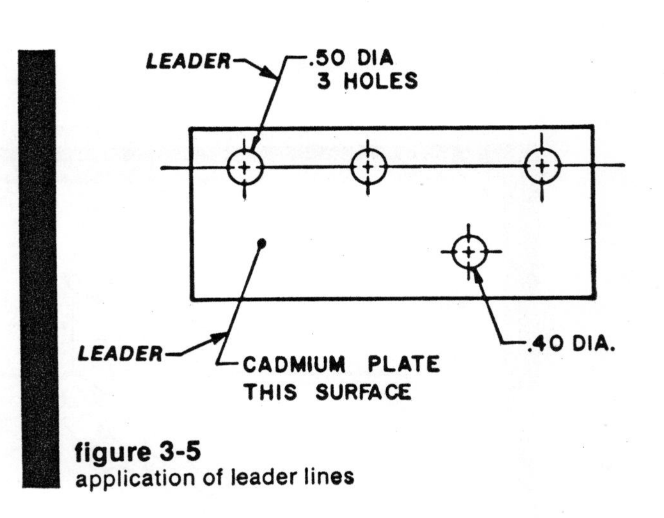

figure 3-5
application of leader lines

leader lines

Leader lines or *leaders* point out various features of an object. Normally, leader lines consist of a small horizontal portion (the shoulder) plus an inclined portion leading to the object. Leaders terminate (end) in arrowheads if they touch an edge of an object. They terminate in a dot if they touch the interior or surface of an object. See Figure 3-5.

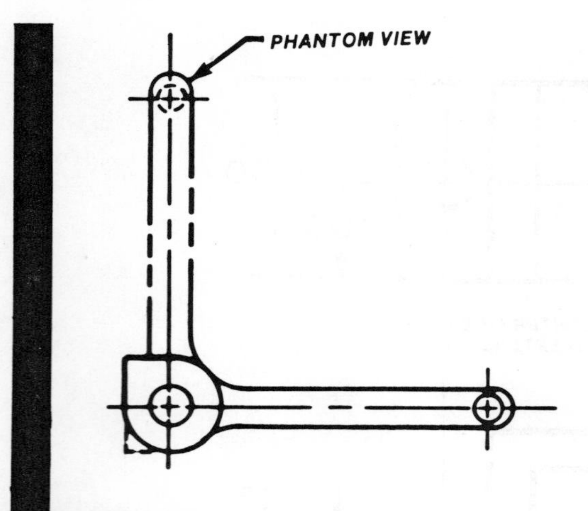

figure 3-6
application of phantom lines for an alternate position

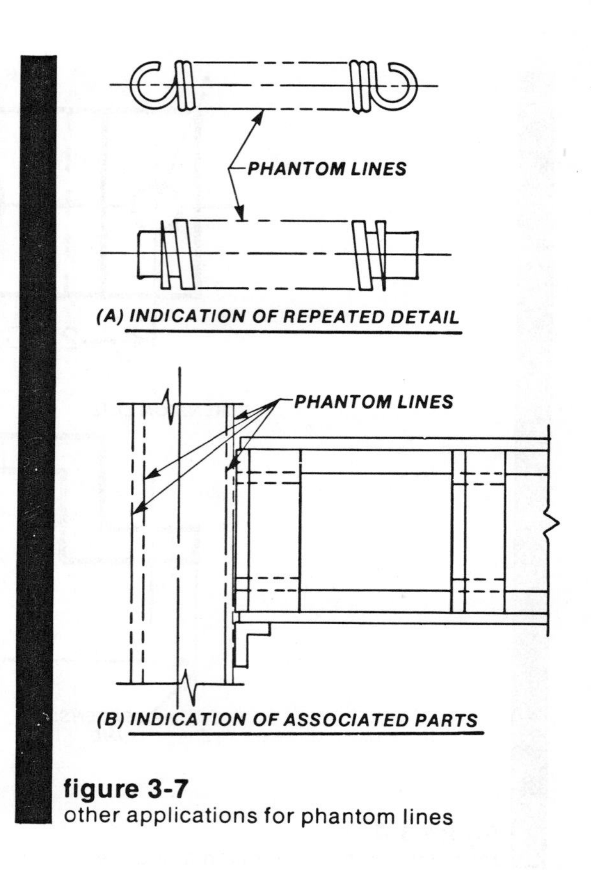

figure 3-7
other applications for phantom lines

phantom lines

Sometimes an object is shown in more than one position to indicate how it operates. This is done with *phantom lines,* alternating long fine lines with double dashes (Figure 3-6). The alternate position is drawn in phantom. It is then called a *phantom view.* Phantom lines also indicate repeated detail (Figure 3-7A) or the position of an associated part (Figure 3-7B).

self-check quiz 3-b

In the view below, identify each of the lines:

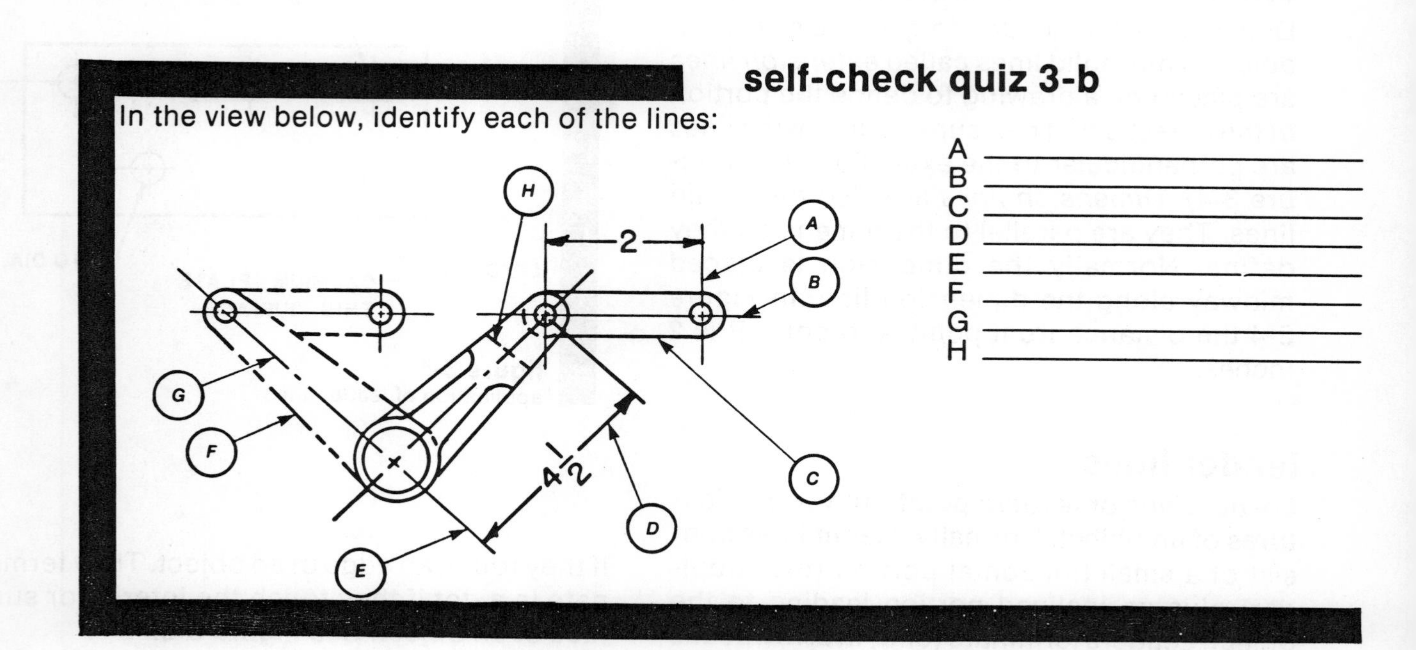

A ______________________
B ______________________
C ______________________
D ______________________
E ______________________
F ______________________
G ______________________
H ______________________

cutting plane lines

Cutting plane lines are similar to phantom lines, but heavier (——— — — ———). Cutting plane lines show the location of cutting planes for section views. Some industries (such as the automotive industry) use a series of heavy, medium-length dashes (— — — — —) for cutting plane lines. Figure 3-8 shows an application of a cutting plane line. A cutting plane line shows where a cross-cut or section is made through an object.

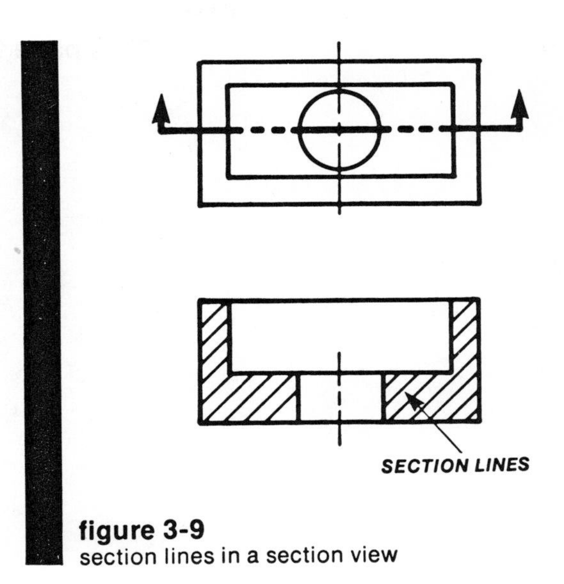

figure 3-9
section lines in a section view

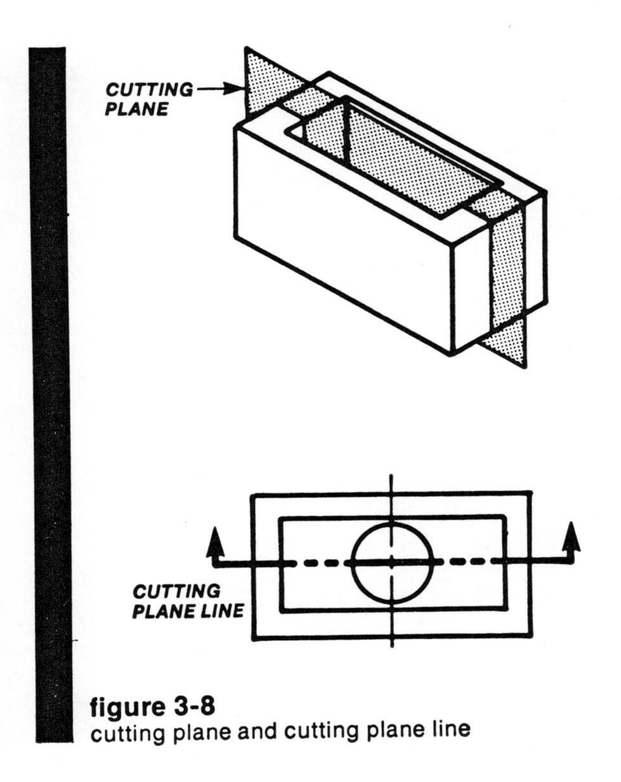

figure 3-8
cutting plane and cutting plane line

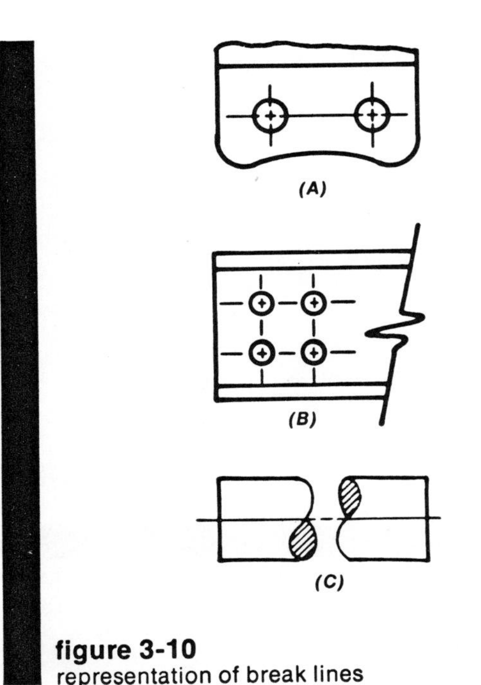

figure 3-10
representation of break lines

section lines

Section lines are used to indicate imaginary cut surfaces of a section view. Figure 3-9 shows a section view with section lines. The section shows the cross-cut made by the cutting plane. The arrows on the cutting plane line point in the direction of viewing the section view.

break lines

Break lines are used when part of the object has been omitted or left out (Figure 3-10A). Break lines are heavy, wavy or jagged lines for short breaks (〜〜〜). Break lines show high, wavy peaks (——⁄\\——) for long breaks (Figure 3-10B). For cylindrical objects break lines give the appearance of "roundness" as shown in Figure 3-10C.

This page for student notes

exercise 3-1

______________________ name

Match each of the lines of the view below with appropriate description.

1 ______
2 ______
3 ______
4 ______
5 ______
6 ______
7 ______
8 ______
9 ______
10 ______

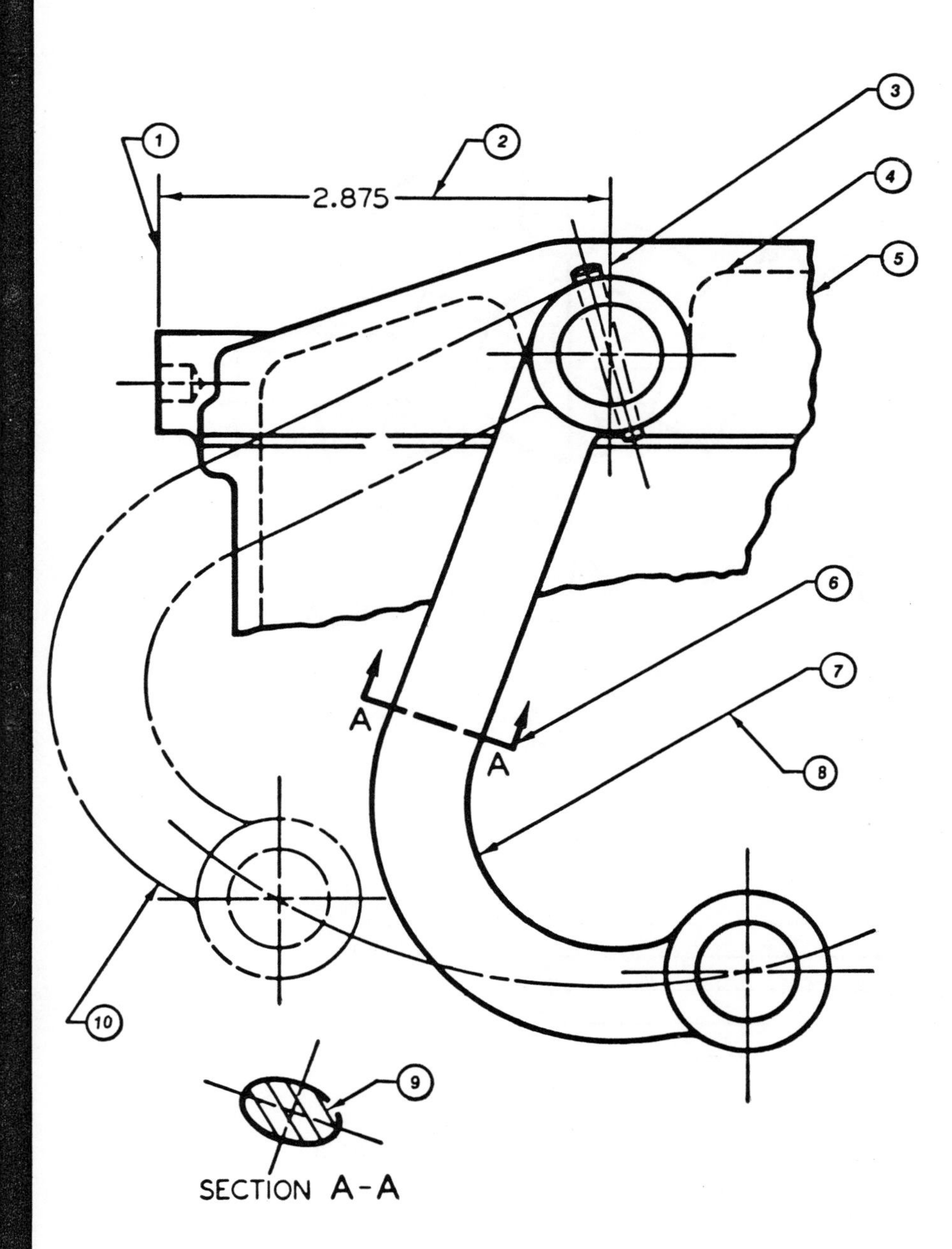

A. Object line
B. Hidden line
C. Center line
D. Extension line
E. Dimension line
F. Section line
G. Phantom line
H. Cutting plane line
I. Leader line
J. Break line

exercise 3-2

name ______________________

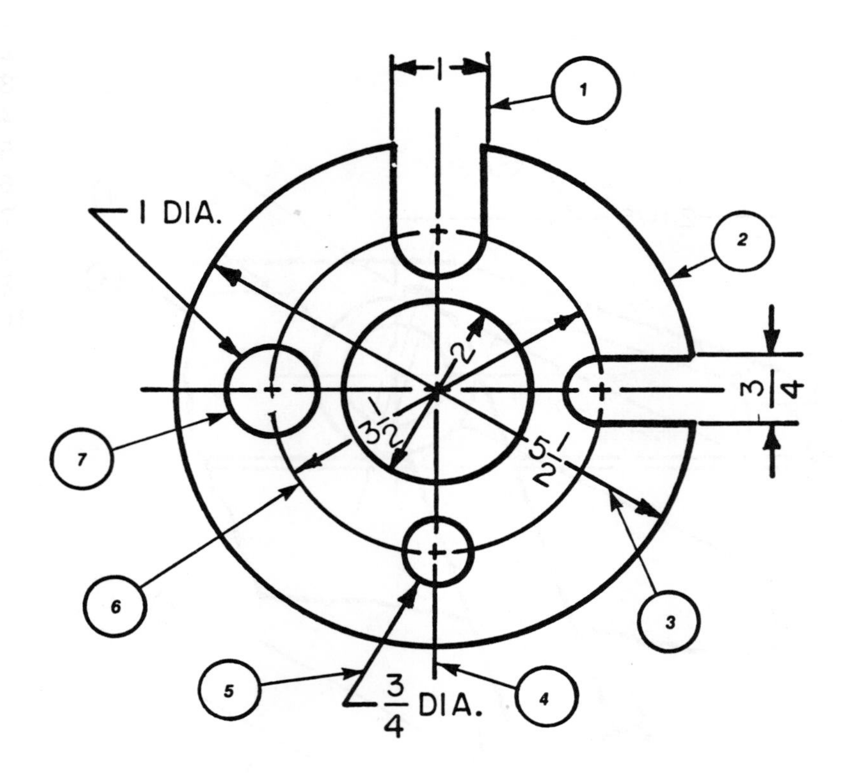

Identify each of the lines noted on the spacer.

1. ______________________
2. ______________________
3. ______________________
4. ______________________
5. ______________________
6. ______________________
7. ______________________
8. What is the overall diameter (distance across) of the spacer?

   ____

9. What is the center hole diameter?

   ____

10. What is the distance from the center of the spacer to the center of the 3/4 DIA. hole?

    ____

orthographic projection

Unit 4

Although a picture might give you an idea of how a part should look, it does not give enough information to machine the part. You need a *working drawing,* which defines the size and the shape of the part. To make a working drawing, a drafter uses *orthographic projection* to draw the part. After completing this unit, you will be able to understand simple working drawings.

third angle orthographic projection

Orthographic projection involves projecting views of an object straight out from the object onto a surface (Figure 4-1). Projection lines are used to project the surface of the object out to the viewing plane.

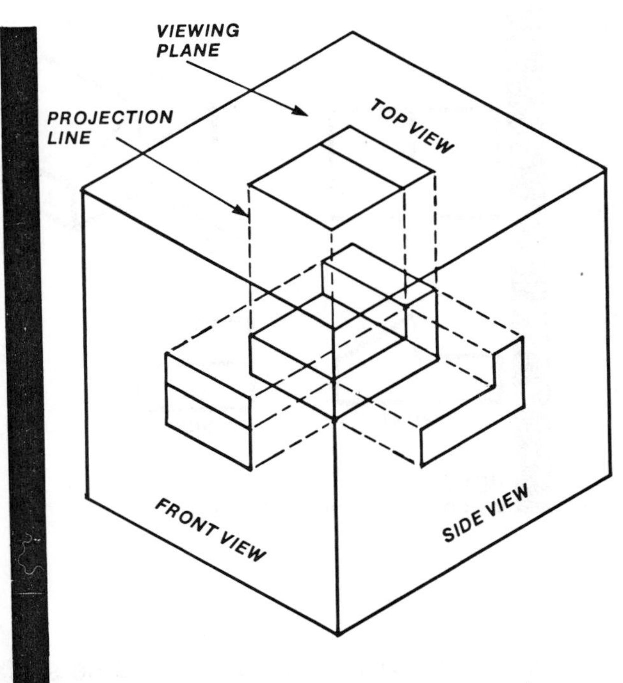

figure 4-1
projection of views in orthographic projection. An imaginary box with transparent sides is used. The views project on the sides of the box.

In the United States, we use *third angle orthographic projection* as shown in Figure 4-2. Some countries use first angle orthographic projection, which is simply a different way of viewing the object. We will consider only third angle orthographic projection in this Unit.

Figure 4-1 shows the three basic views of an object viewed directly from the front, top and right sides. We project these views onto the surfaces of an imaginary transparent box.

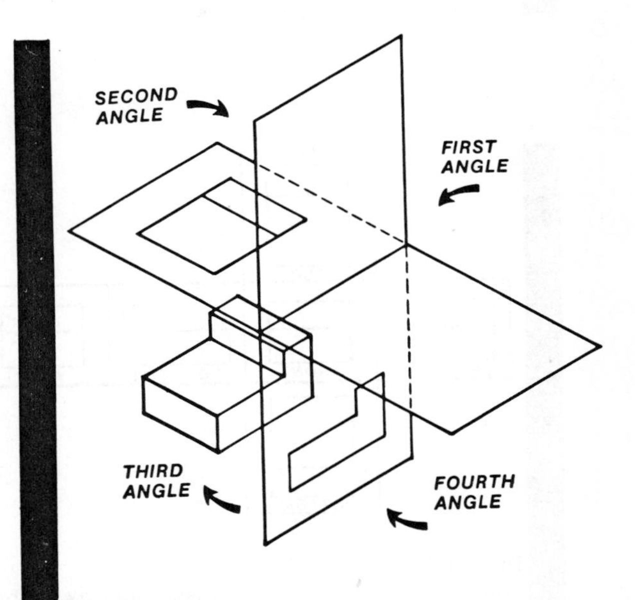

figure 4-2
an object shown in *third angle orthographic* projection

Other views, such as the rear, bottom and left side views (Figure 4-3), can also be projected onto the box. We arrange the views on a drawing as though the box were unfolded around the front view as shown in Figure 4-4.

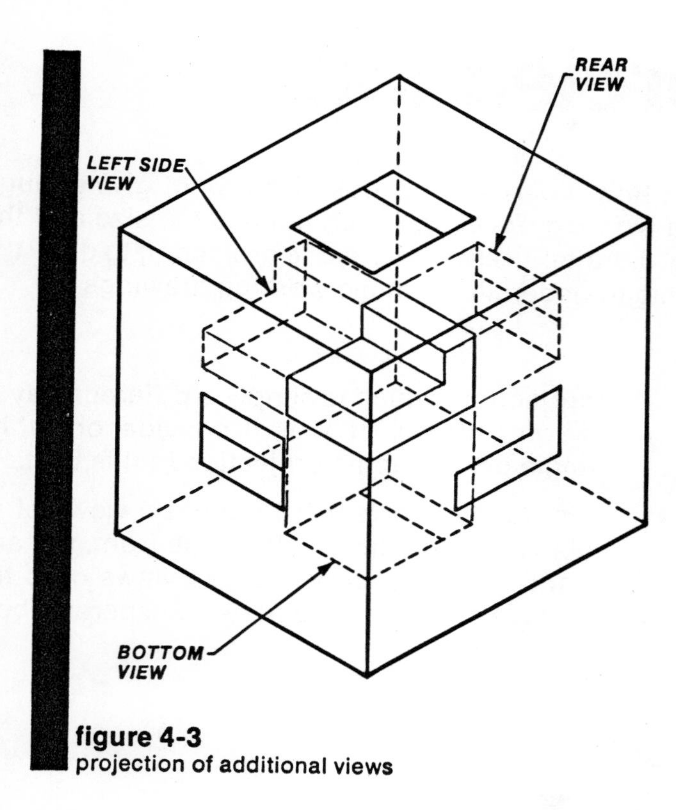

figure 4-3
projection of additional views

figure 4-5
viewing an object by its principal views (front, top and right side)

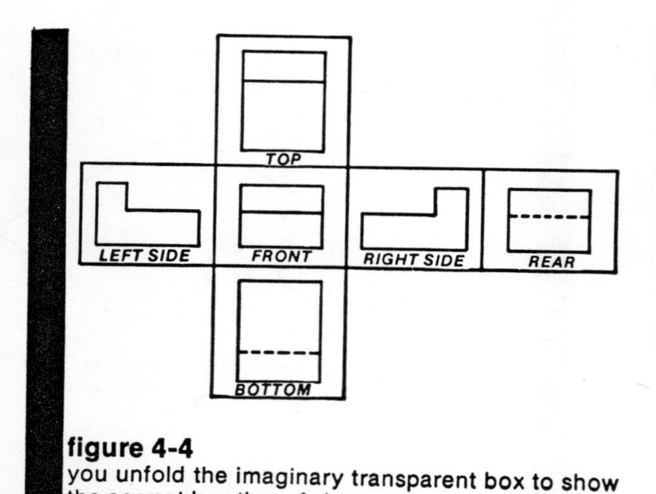

figure 4-4
you unfold the imaginary transparent box to show the correct location of views

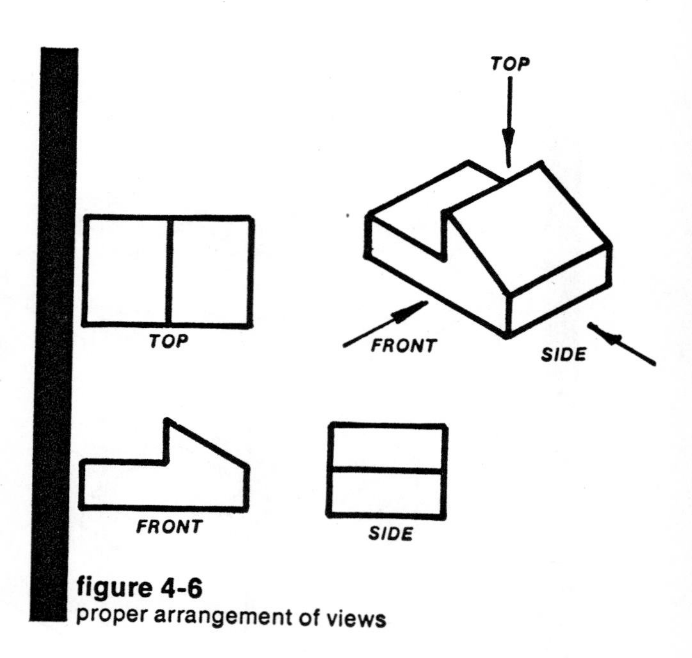

figure 4-6
proper arrangement of views

In order to undertand third angle orthographic projection better, consider the object as stationary (sitting still) while you move around it (Figure 4-5). First, view the object directly from the front. Then view it from the top; then from the right side. All surfaces appear as surfaces or edges. Some surfaces may be hidden in a particular view.

Study the object shown in Figure 4-6. See if you can visualize projecting the views of the object onto the transparent box. Then visualize unfolding the box to display the views on the flat drawing surface.

In the example of Figure 4-7, observe how the views are arranged. Notice how surfaces and

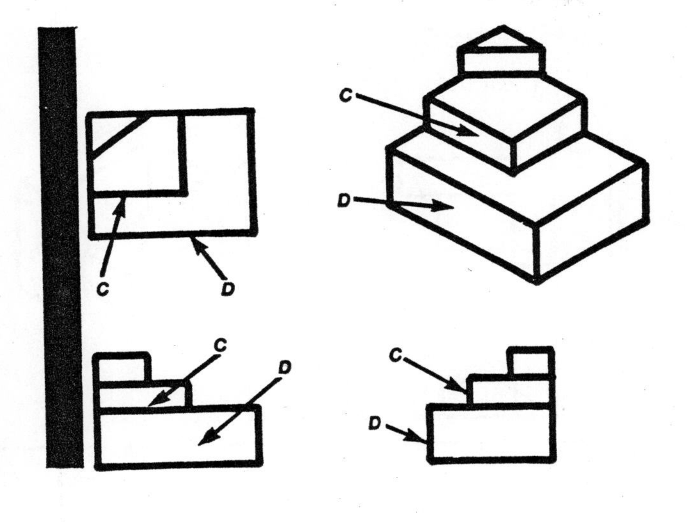

figure 4-7
vertical surfaces of the front view appear as edges in top and side views

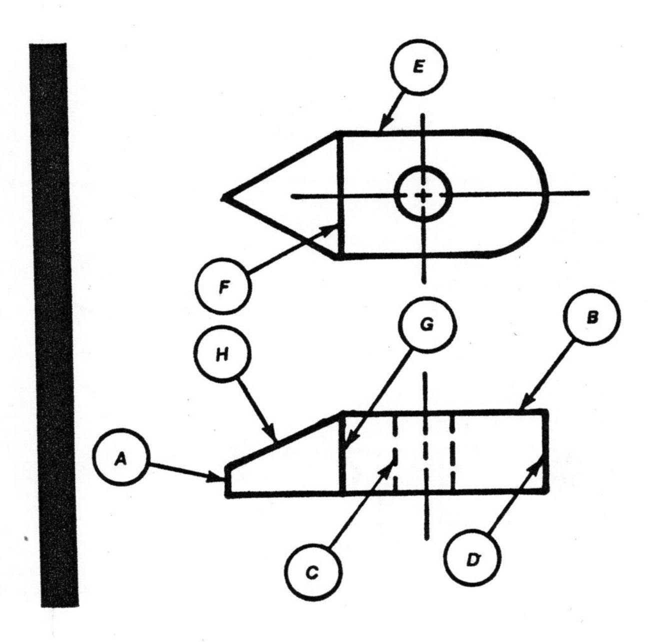

figure 4-8
lines represent edge views, intersections, and contours

edges line up from one view to another. Notice that vertical surfaces in the front view (surfaces C and D) appear as edges represented by single lines in the top and side views.

Lines in a view can represent many things. In Figure 4-8, lines B, H, and E represent edge views of surfaces. Lines A, F, and G represent intersections of surfaces. Lines C and D represent the contours of curved surfaces.

self-check quiz 4-a

Identify the surfaces in the orthographic views. FV indicates front view.

EXAMPLE

A B C FV

1

A B C D E F G H FV

2

A B C D E FV

3

A B C D E F G H FV

4

A B C D E FV

5

A B C D FV

6

A B C FV

7

A B C D E F FV

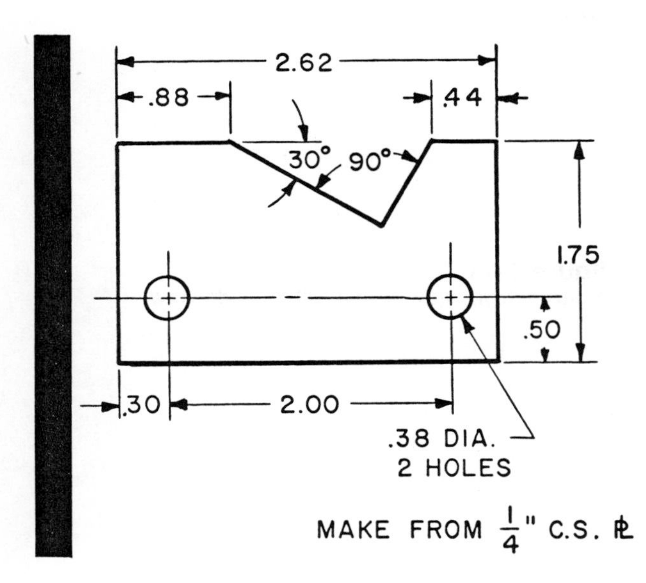

figure 4-9
a single-view drawing

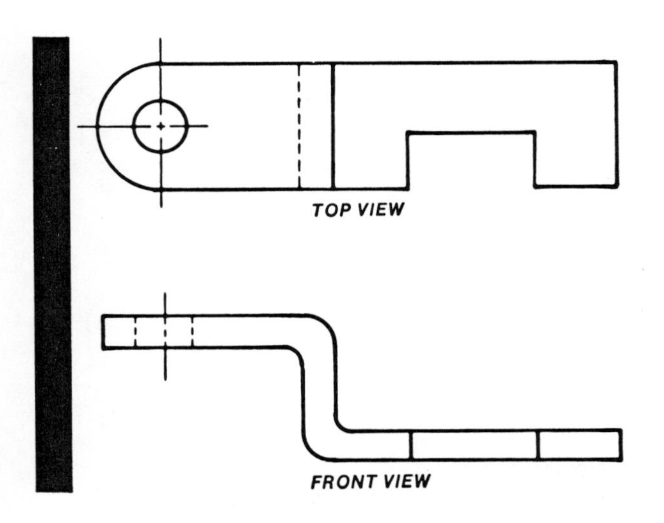

figure 4-10
a two-view drawing

necessary views and arrangements

The drafter usually draws only those views and details which are necessary to describe the object. A single view may do the job. Additional information is given in a note (Figure 4-9). Often, two views are enough (Figure 4-10). Sometimes a partial view gives an adequate description. (Figure 4-11).

The drafter chooses the best view or the largest view as the front view. Then a top or side view (or both) is added. That side view which shows more visible edges is usually chosen. Section views and other details are added as necessary to describe the object. The side view is projected off the front or off the top view (Figure 4-12), depending on the drafter's preference. Normally it is projected off the front view.

SYMMETRICAL
ABOUT ℄

figure 4-11
a partial front view with a full left side view

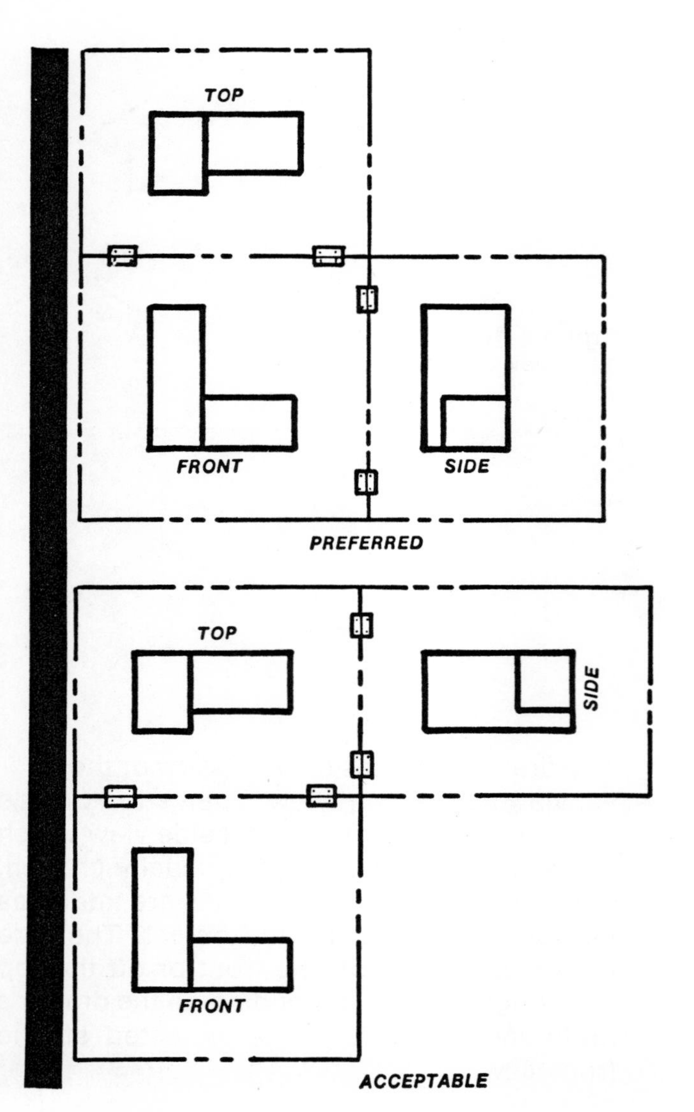

figure 4-12
alternate locations for the side view

self-check quiz 4-b

Identify each point on the orthographic views from the corresponding point on the pictorial view.

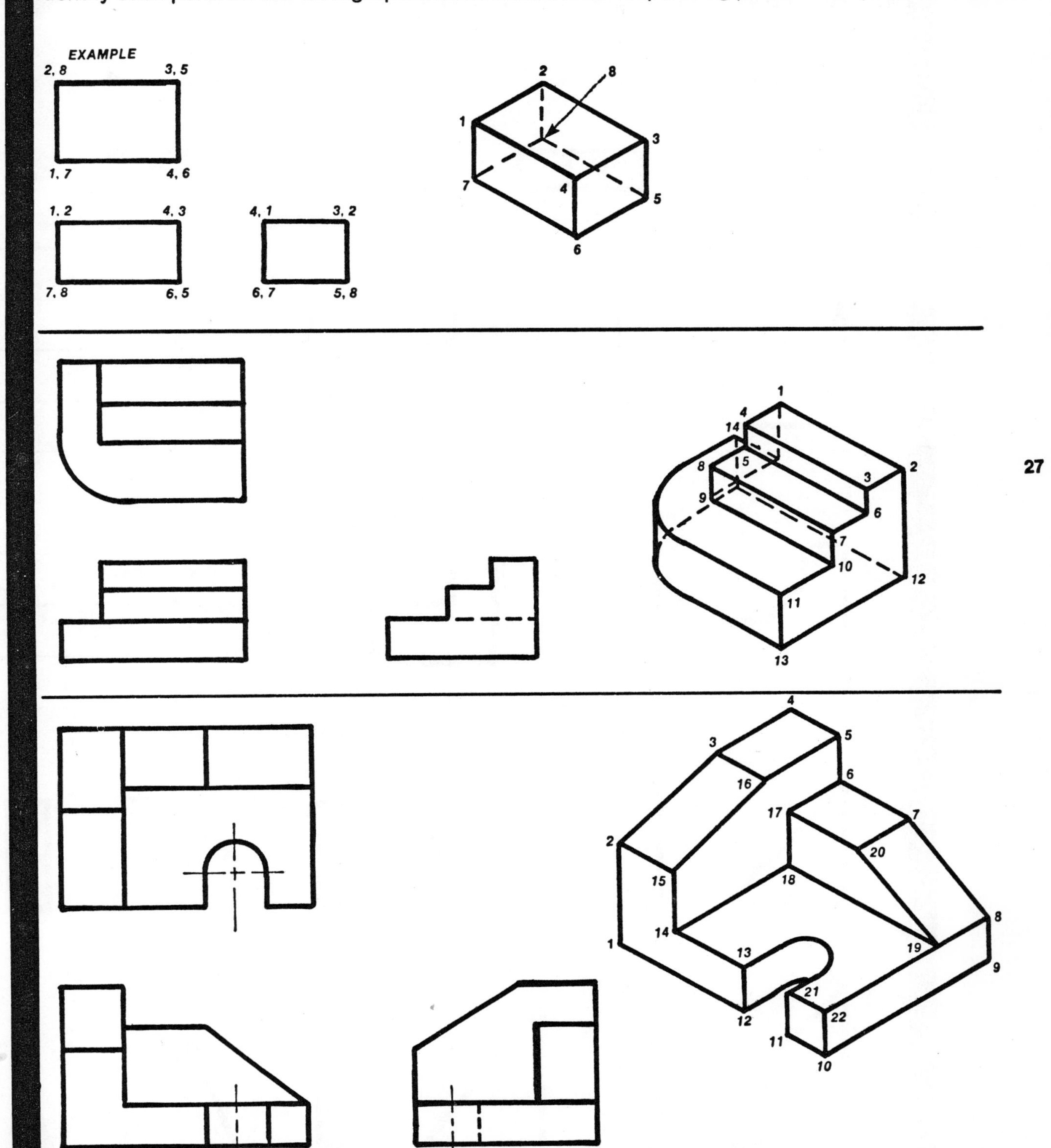

exercise 4-1

______________________________ name

Match the top views to the objects below. Top views may be used more than once.

A B C D

E F G H

1. ______________
2. ______________
3. ______________
4. ______________
5. ______________
6. ______________
7. ______________
8. ______________
9. ______________
10. ______________

1 2 3 4 5 6 7 8 9 10

exercise 4-2

_______________________________ name

1 Match the lettered surface of the pictorial view with the corresponding numbered surface in each orthographic view.

SURFACE	TOP VIEW	FRONT VIEW	SIDE VIEW
A			
B			
C			
D			
E			
F		—	
G			
H			
I			
J			
K			
L			
—	M		
—	N		
—	O		
—	P		
—	Q		
—	R		
—	S		

2. Match the lettered surface of the top view with the corresponding surface in the front and side views.

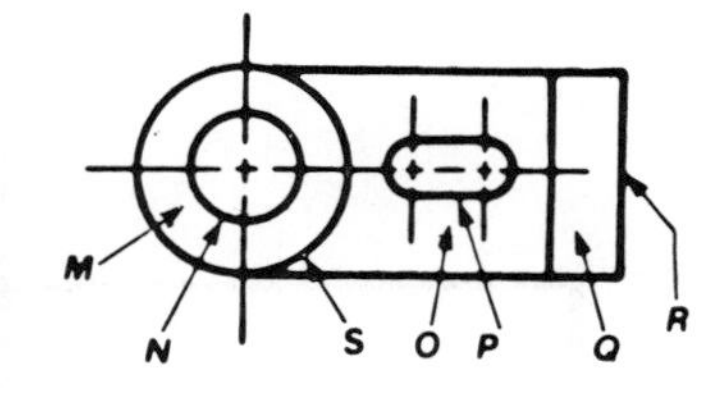

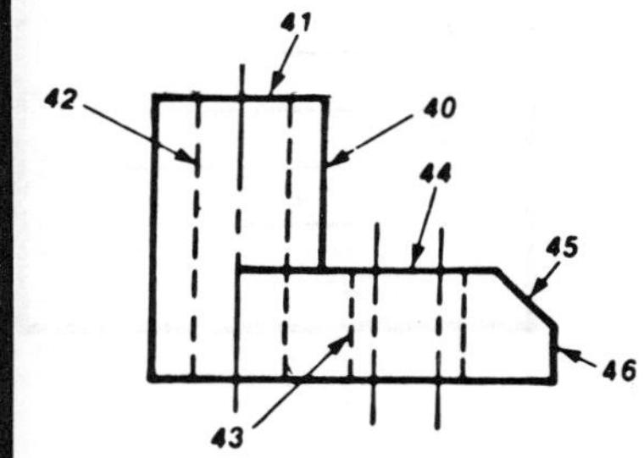

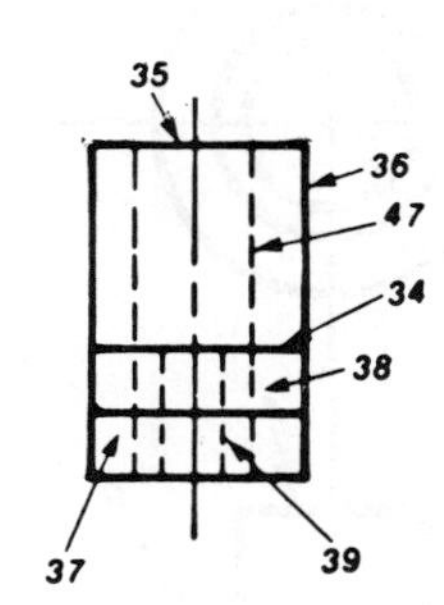

exercise 4-3

______________________________ name

Match the pegs of the front and side views of the object below.

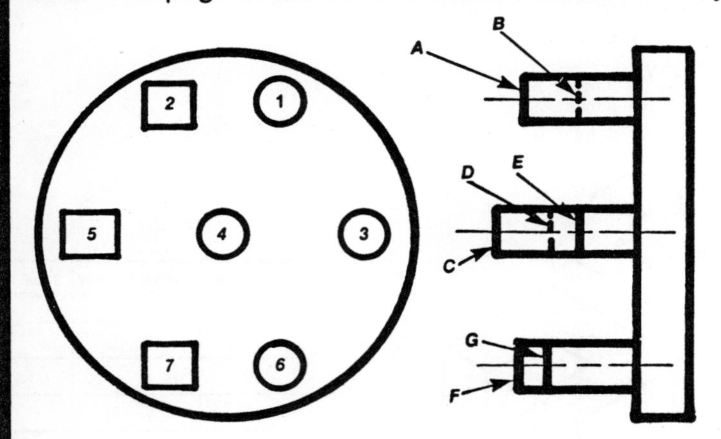

1. ____________
2. ____________
3. ____________
4. ____________
5. ____________
6. ____________
7. ____________

Match the shaded surface in the top view with the corresponding surface in the front view for each object below:

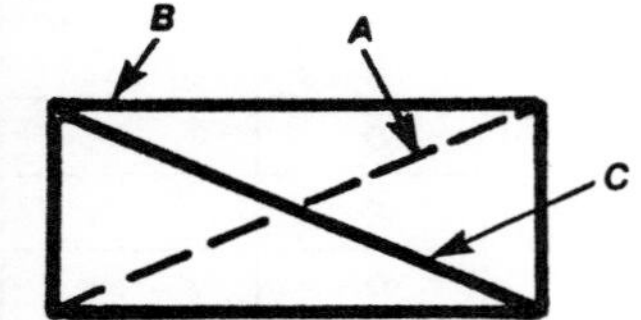

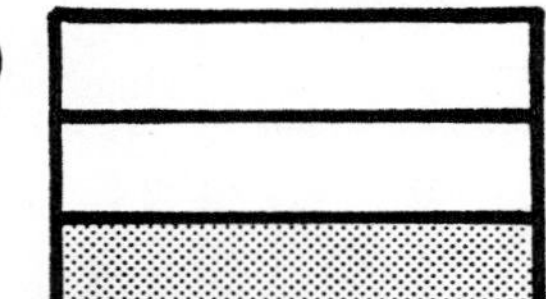

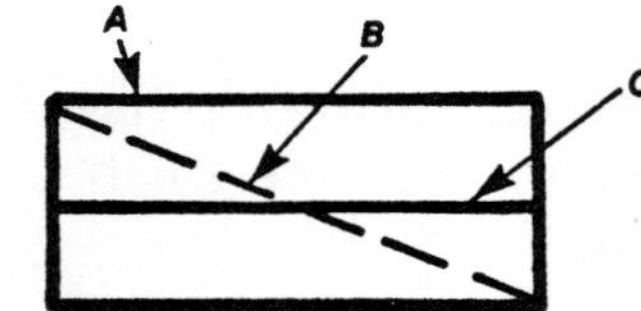

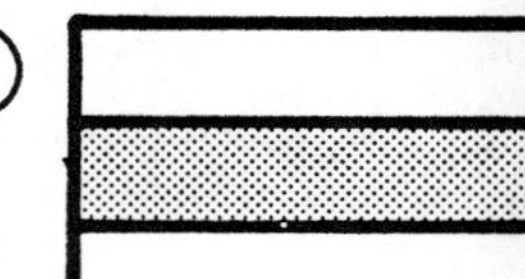

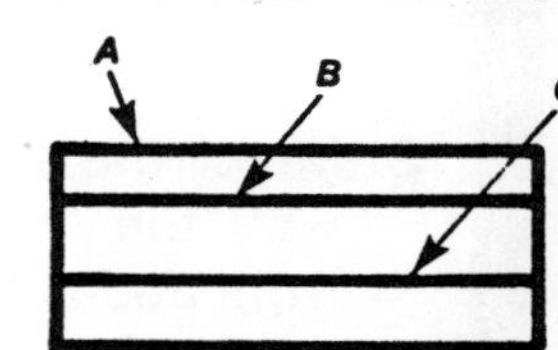

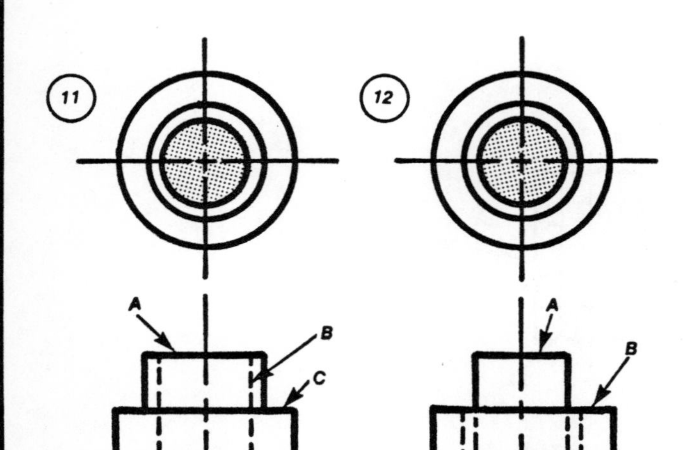

8. ____________
9. ____________
10. ____________
11. ____________
12. ____________

exercise 4-4

______________________________ name

Sketch front, top, and right side orthographic views of the objects shown. Each mark on the objects represents ¼ inch. Use a separate piece of paper.

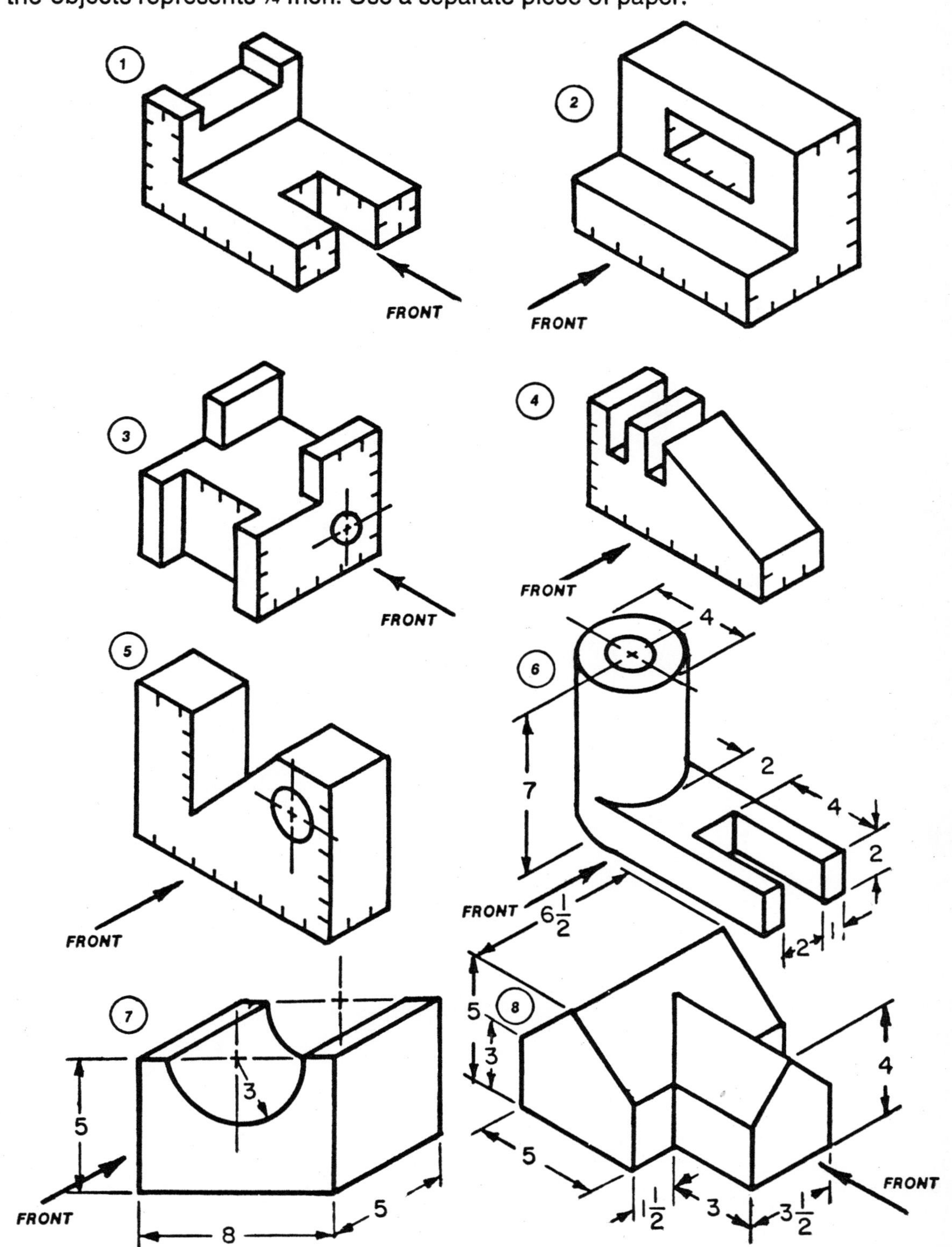

This page for student notes

freehand sketching

In this unit you will develop the skill of sketching objects freehand.

communicating on the job

Sometimes you communicate an idea better with sketches than with words. *Sketches* are rough, freehand drawings. Sketches are the first step in solving a design problem. They are helpful when drawings do not exist for a worn-out or broken machine part. Sketches are an effective way to express a proposed design improvement or new idea. If you develop the technique of freehand sketching, you will be able to communicate your design ideas on the job. Sketching is not difficult if you follow basic rules.

sketching lines freehand

For sketching, use a medium or soft pencil. (Use a #2 pencil if H, F or HB pencils are not available.) A long conical point does the best job. If thick lines are needed, the point should be slightly rounded. Hold the pencil with a comfortable, firm grip as you would in normal handwriting (Figure 5-1). Loosen up your arm and hand muscles before you start sketching. As you sketch, rotate your pencil between strokes so that the pencil point stays sharp.

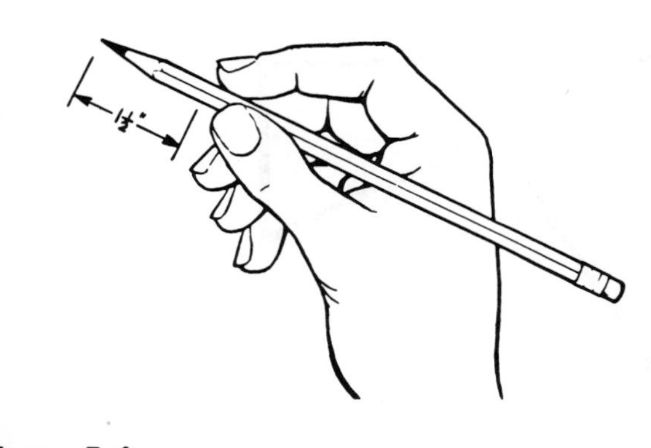

figure 5-1
the proper hand position for sketching

sketching horizontal lines

A horizontal line is drawn with a series of short strokes (Figure 5-2). Move the entire forearm with the pencil. Keep your eyes on the target (an end point located across the paper) and not on the pencil. Sketch horizontal lines from left to right. Figure 5-3 shows the steps in sketching a horizontal line.

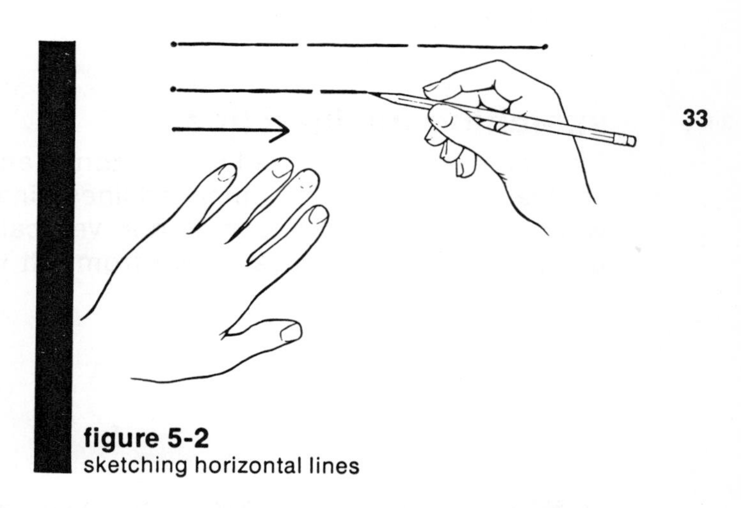

figure 5-2
sketching horizontal lines

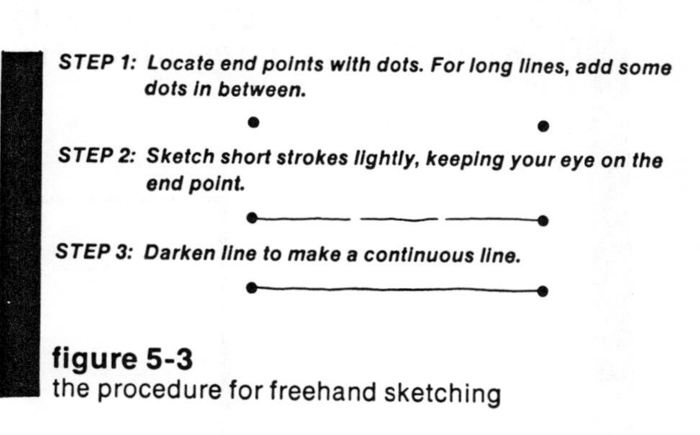

figure 5-3
the procedure for freehand sketching

sketching vertical lines

A vertical line is drawn in a similar fashion to a horizontal line, only from top to bottom. A

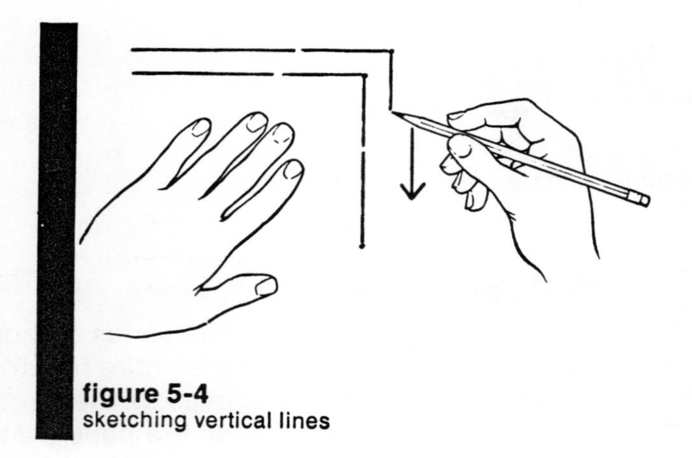
figure 5-4
sketching vertical lines

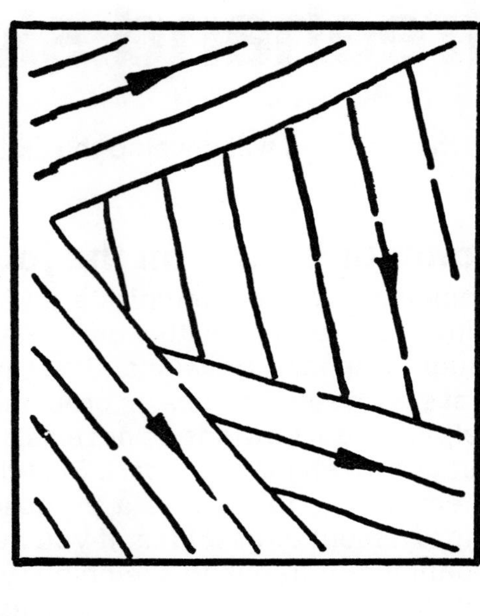
figure 5-5
sketching inclined lines

"fingers and base of hand" movement is used, moving the entire forearm after a few strokes have been made (Figure 5-4).

sketching inclined lines

The same techniques used for horizontal and vertical lines are used for inclined lines (lines which are neither horizontal nor vertical). Inclined lines are always drawn from left to right (Figure 5-5). Sometimes it is easier to turn the paper and sketch the line as you would a horizontal line.

self-check quiz 5-a

Draw an arrowhead on each line of this figure to indicate in which direction the line should be sketched.

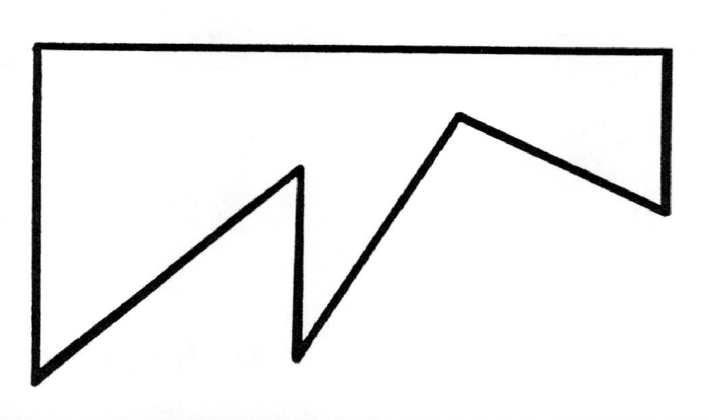

sketching arcs and circles

One method for sketching arcs and circles is the *triangle technique.* The triangle technique is illustrated in the following procedure for sketching a circle:

1. Draw center lines for the circle (Figure 5-6A).
2. Mark the radius of the circle on each center line (Figure 5-6B). (The radius is the distance from the center of the circle to the edge of the circle.)
3. Block in the circle with a square through the radius points of step 2 (Figure 5-6C).
4. Connect the radius points of step 2 so that triangles are formed, and locate the center of each triangle with a mark (Figure 5-6D).
5. Sketch short arcs through the radius points and the triangle centers (Figure 5-6E). Draw the arcs with clockwise strokes.
6. Darken the arcs to make a continuous circle. Erase all construction lines if they interfere with the completed sketch (Figure 5-6F).

Since arcs are simply portions of a circle, they are sketched in a similar manner as outlined above. See Figure 5-7.

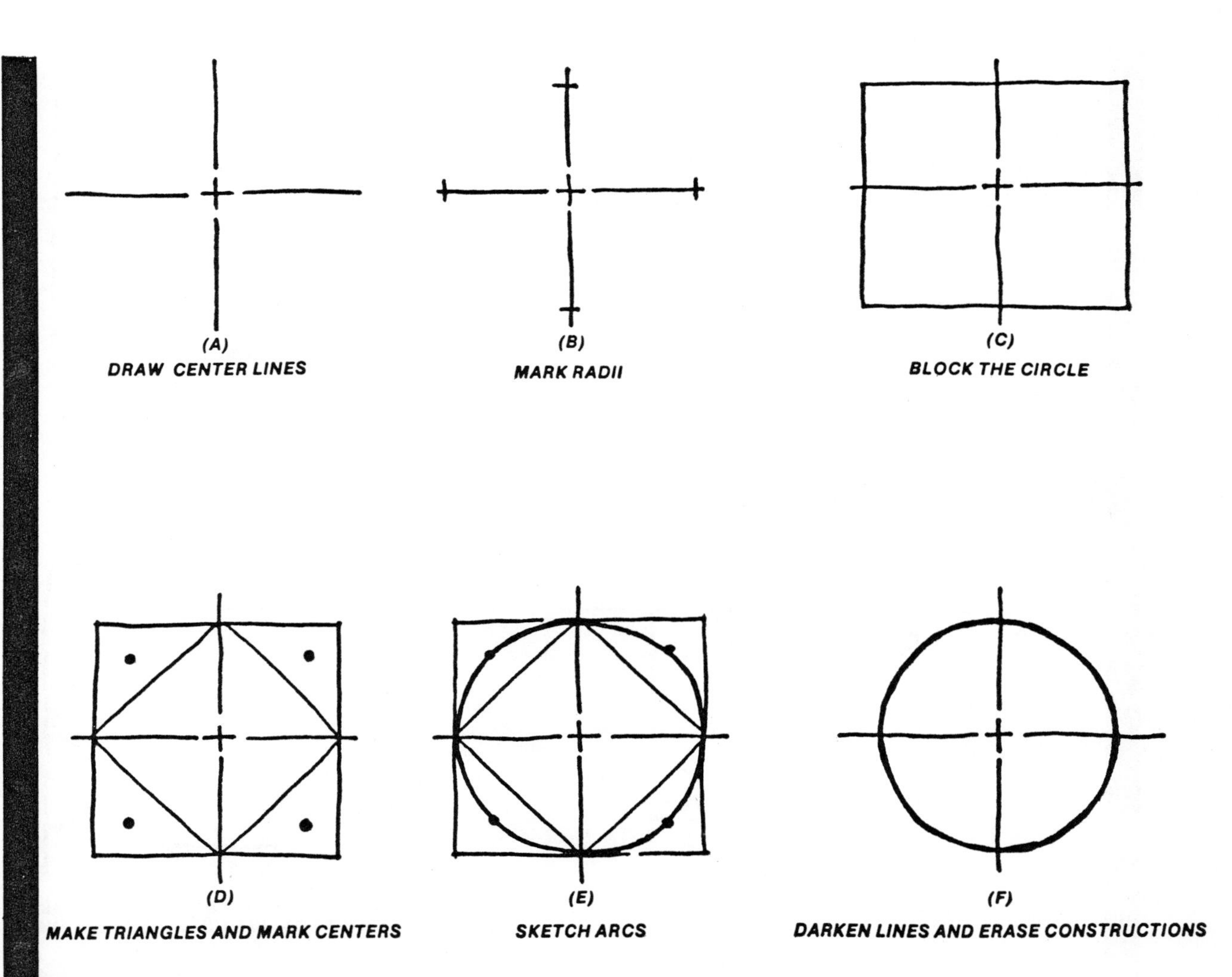

figure 5-6
procedure for sketching a circle

figure 5-7
sketching an arc

sketching irregular shapes

A complicated object can easily be sketched by first dividing the object into squares and rectangles. Then straight lines, arcs, and circles are used to define the true shape of the object. Figure 5-8 illustrates how this is done.

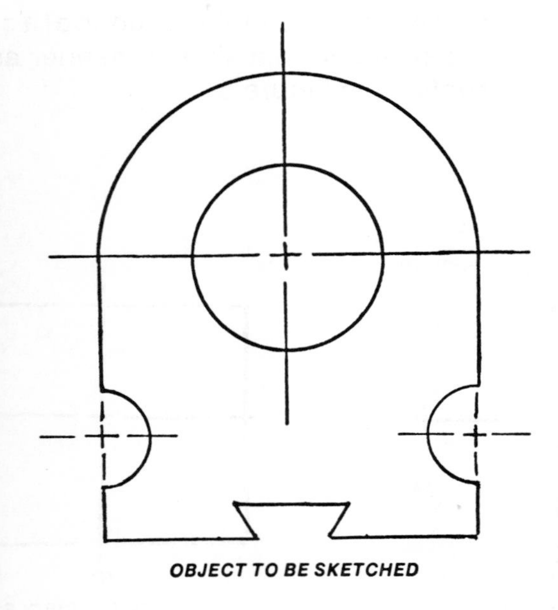

OBJECT TO BE SKETCHED

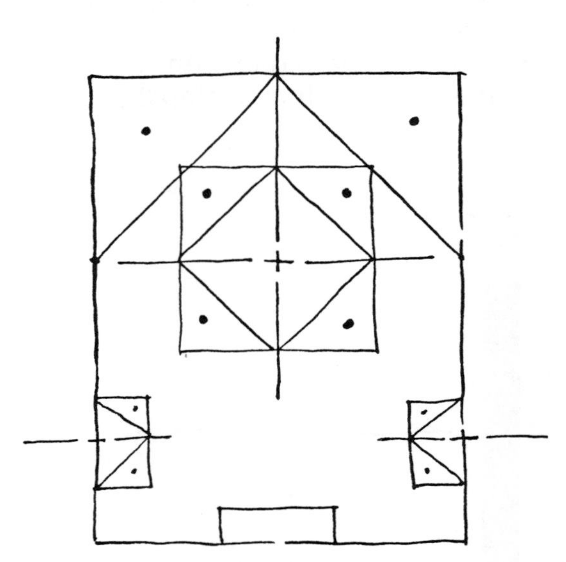

1. BLOCK THE OBJECT AND APPLY THE TRIANGLE TECHNIQUE

2. SKETCH LINES AND ARCS

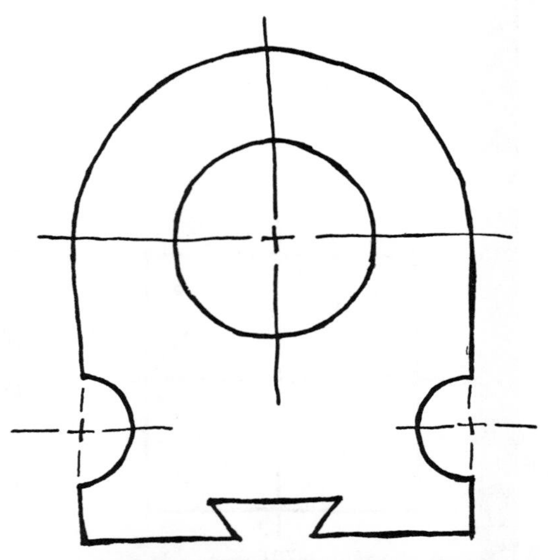

3. DARKEN THE OBJECT LINES AND ERASE OBJECTIONABLE CONSTRUCTION LINES

figure 5-8
the procedure for sketching a complicated object

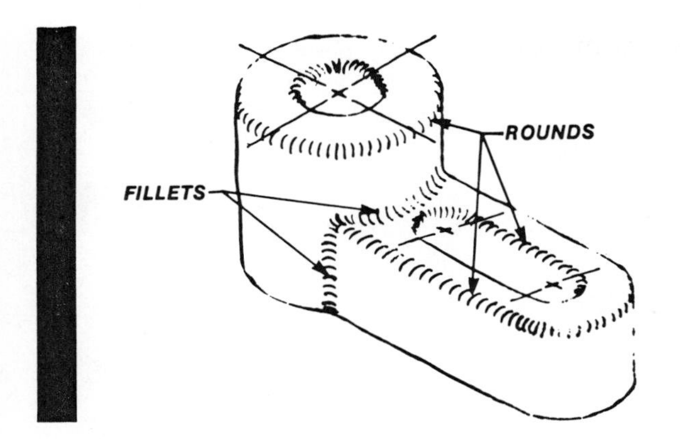

figure 5-9
sketching fillets and rounds

sketching fillets and rounds

Sharp corners and sharp edges on parts are often avoided because they result in higher stress and lower strength of the parts. This is particularly true with castings and formed metal parts. Figure 5-9 shows a part with rounded corners and edges. When an inside corner is rounded, it is called a *fillet*. When an outside edge is rounded, it is called a *round*. Short curved lines represent fillets and rounds in pictorial views.

This page for student notes

exercise 5-1

_______________ name

Sketch lines according to the pattern shown.

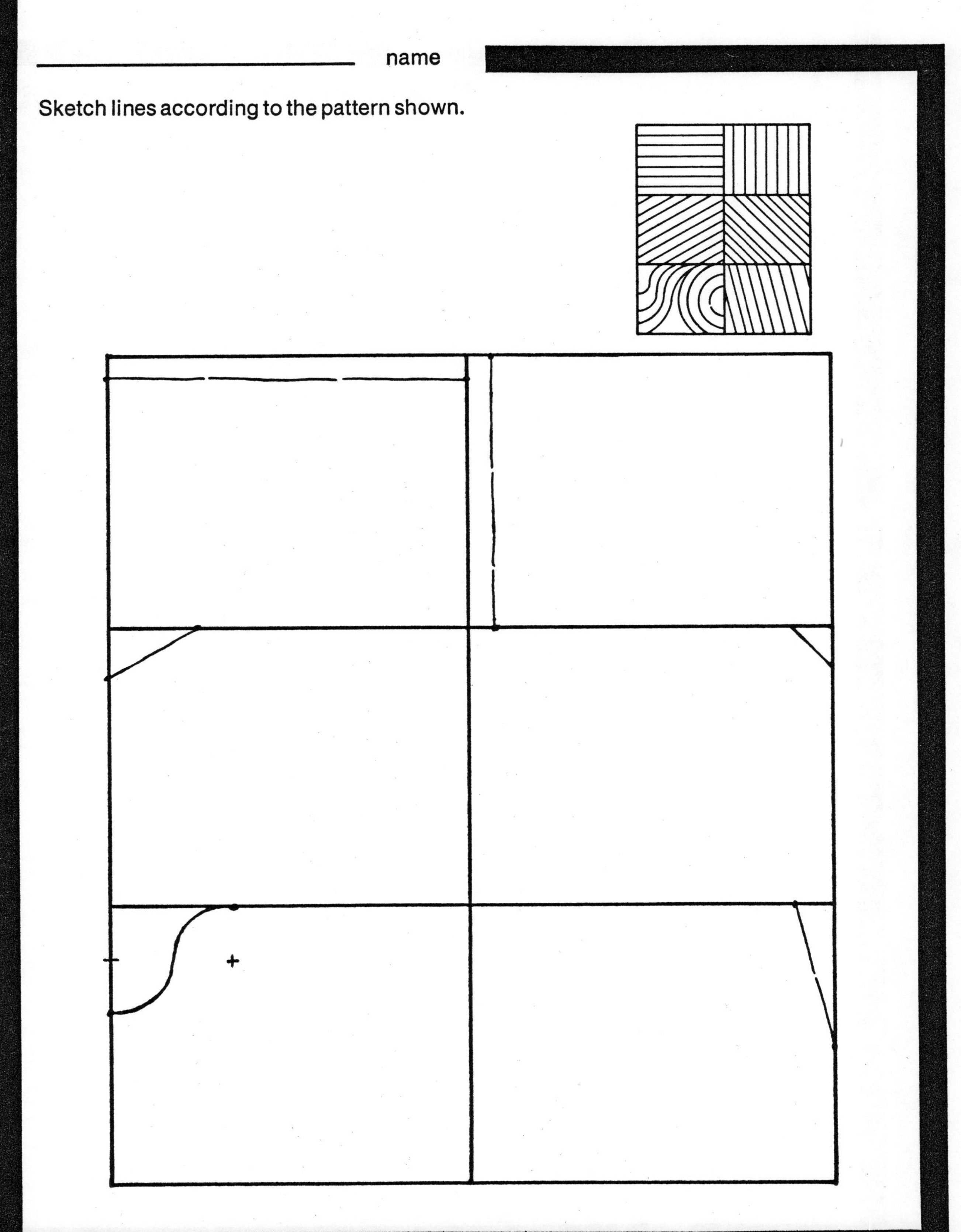

exercise 5-2

______________________ name

Connect the dots below, following the sketching procedures you have learned. Put arrowheads on each line to indicate the correct sketching direction.

exercise 5-3

______________________________ name

Below, sketch the figure shown, according to the sketching procedures you have learned. Put arrowheads on each line to indicate the correct sketching direction.

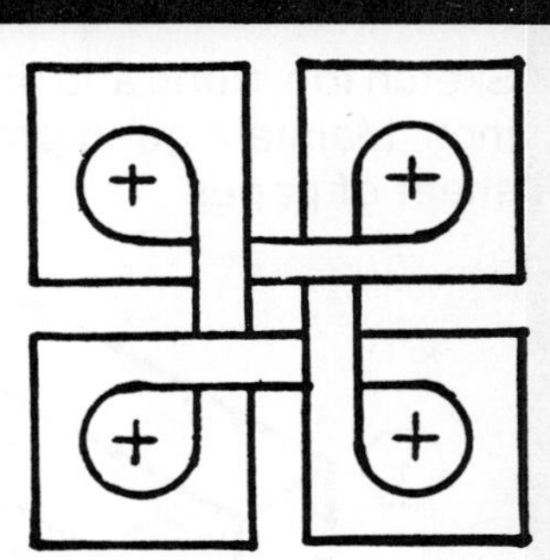

exercise 5-4

_________________________________ name

Sketch top, front, and right side views of the objects shown. Each mark on the objects represents ¼ inch. Maintain good proportion. "FV" indicates the front view of each object. Use a separate sheet of paper.

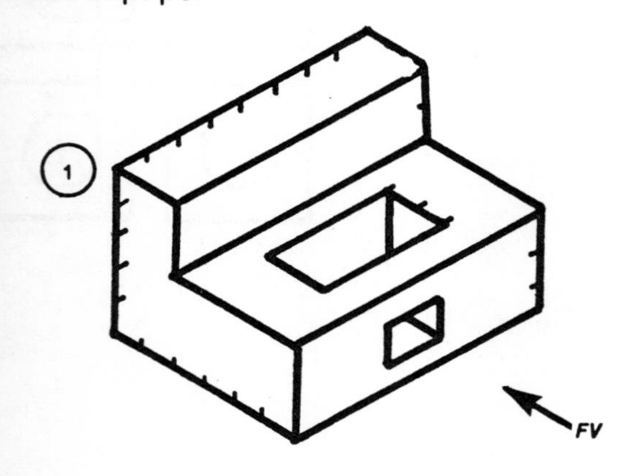

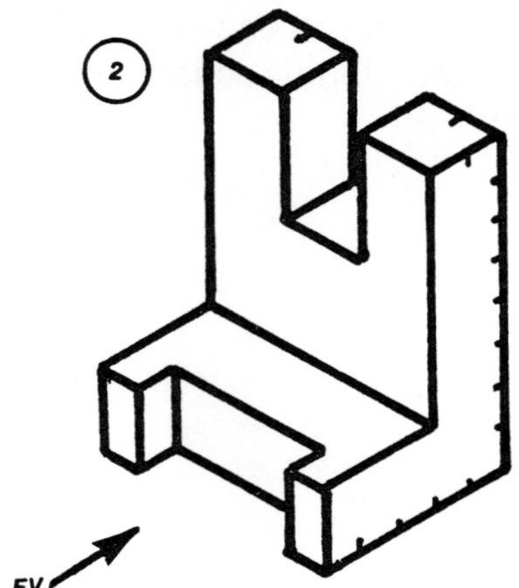

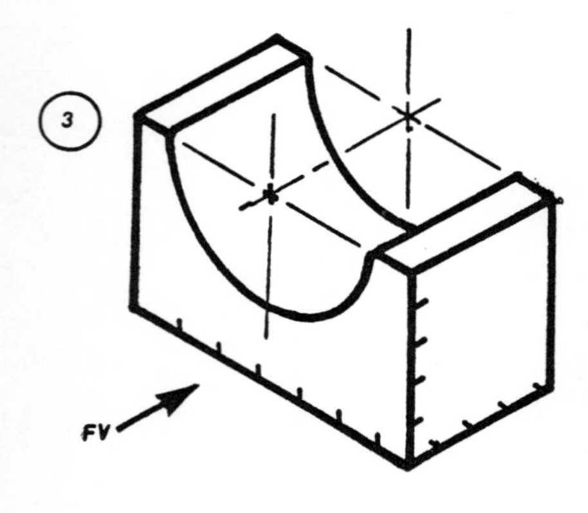

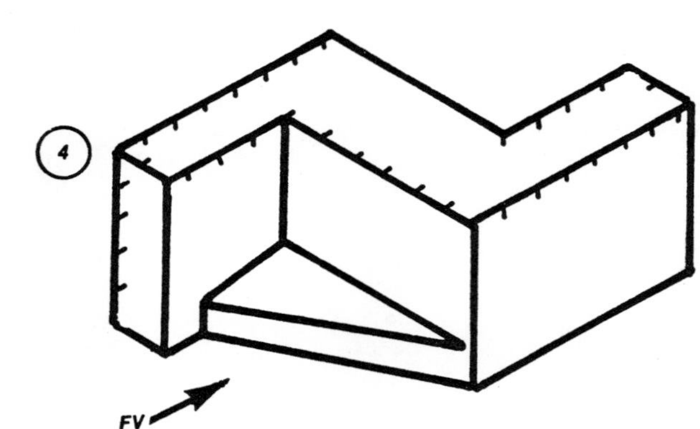

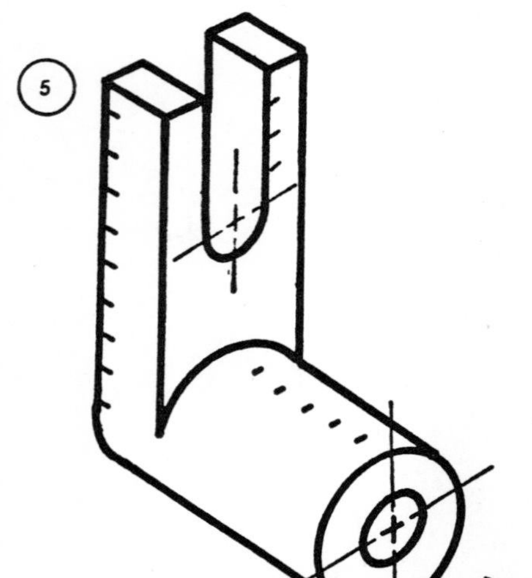

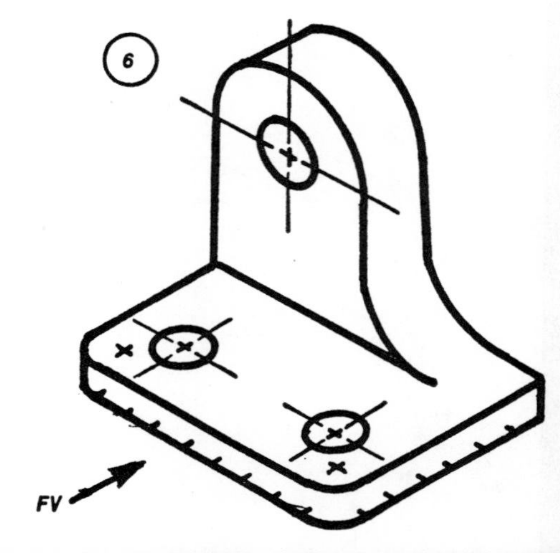

pictorial drawings

Unit 6

Many times a *pictorial* view (a view like a picture or photograph) conveys the necessary information. After completing this unit you will be able to recognize and interpret the different types of pictorial drawings. You will also develop a technique for sketching isometric and oblique pictorial views.

A pictorial drawing gives an easy-reading view of an object. Designers, drafters, architects and engineers use three different types of pictorial views: perspective, isometric and oblique. Figure 6-1 shows an angle bracket sketched by the three different methods.

perspective views

A perspective view shows an object with the sides converging (coming together) in the distance (Figure 6-1). Perspective is used in architectural drafting and in fine art painting. Perspective views are not used in the machine shop.

isometric views

Isometric views are used to show the general appearance of a part or assembly. In an isometric view, all lines parallel on the object are also parallel on the drawing (Figure 6-2). The object is drawn by showing vertical lines of the object in a vertical position. Horizontal lines

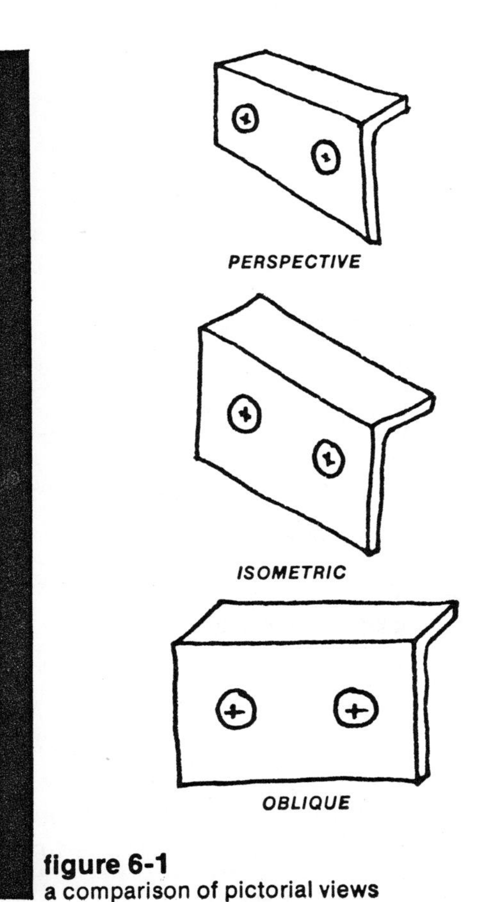

figure 6-1
a comparison of pictorial views

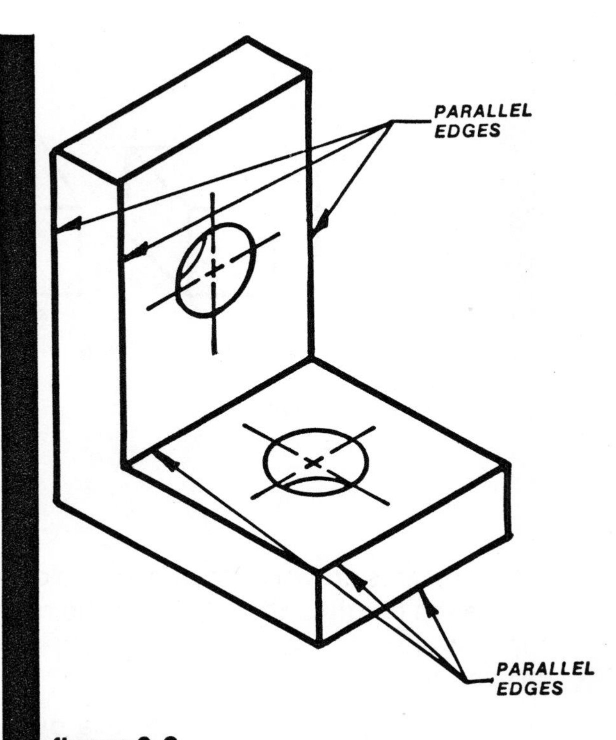

figure 6-2
parallel edges of an object appear parallel in an isometric view of the object

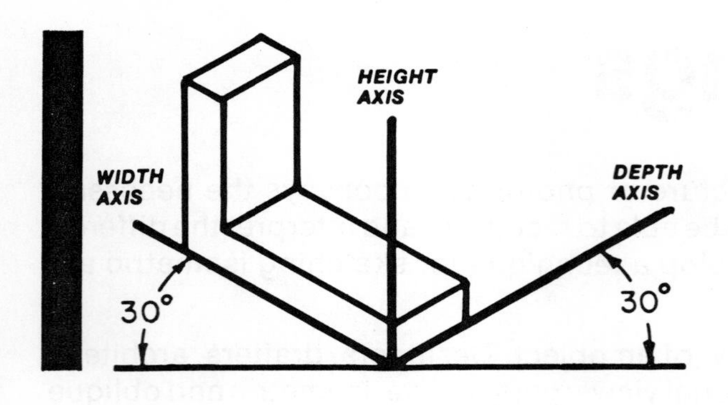

figure 6-3
in isometric views measurements are along the axes

are shown at an angle of 30° to the horizontal. Height is measured true vertical. Width and depth are measured along the 30° axes (Figure 6-3).

oblique views

An oblique view of an object shows the front side true size and shape. The front side is like an orthographic view. The top and side recede at some angle, usually 30°, 45° or 60°, to the horizontal (Figure 6-4).

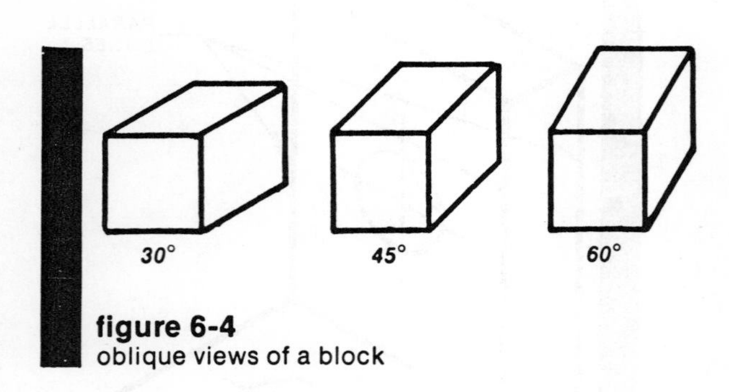

figure 6-4
oblique views of a block

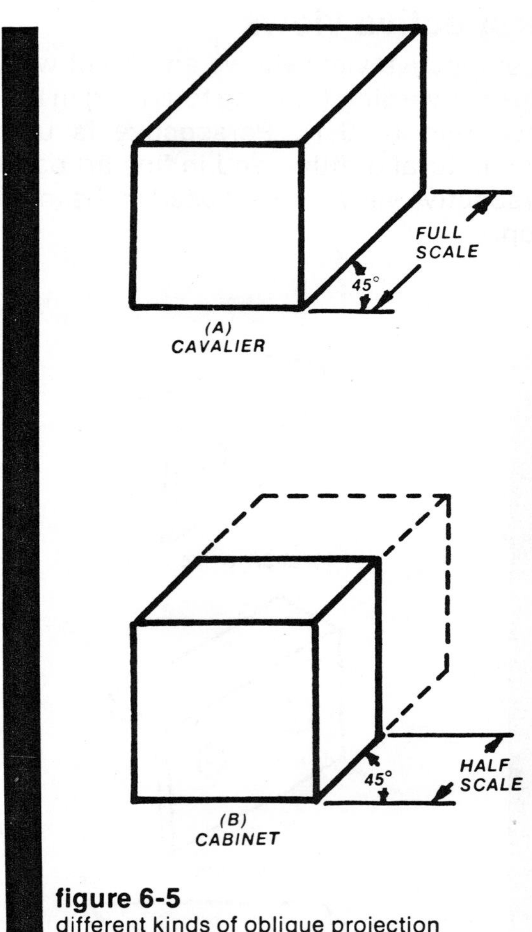

figure 6-5
different kinds of oblique projection

If the object is drawn *full scale* for the depth as well as the height and width, and the receding angle is 45°, the view is known as *cavalier projection* (Figure 6-5A).

Sometimes the receding lines are drawn by using *half scale* measurements to give a more pleasing appearance. This is known as *cabinet projection* (Figure 6-5B). Cabinet drawings are usually drawn with the receding angle of 30° or 45°.

self-check quiz 6-a

Identify each of the following views as isometric, cavalier oblique or cabinet oblique.

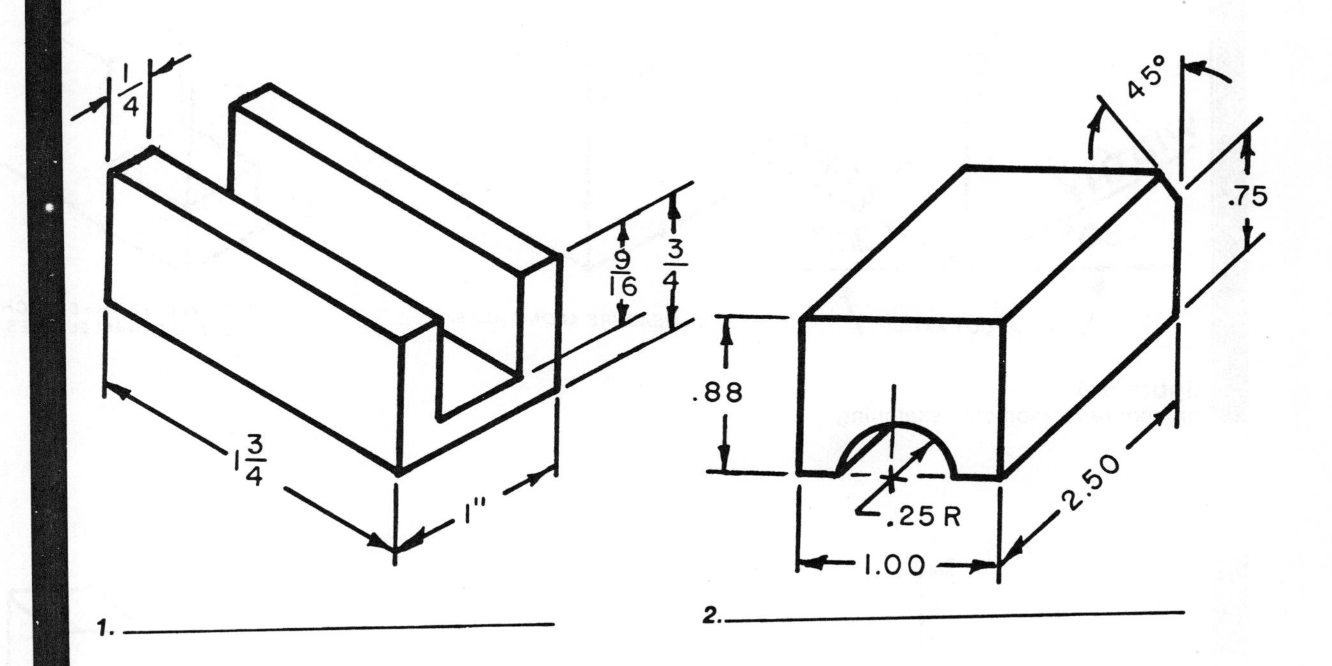

1. ______________________

2. ______________________

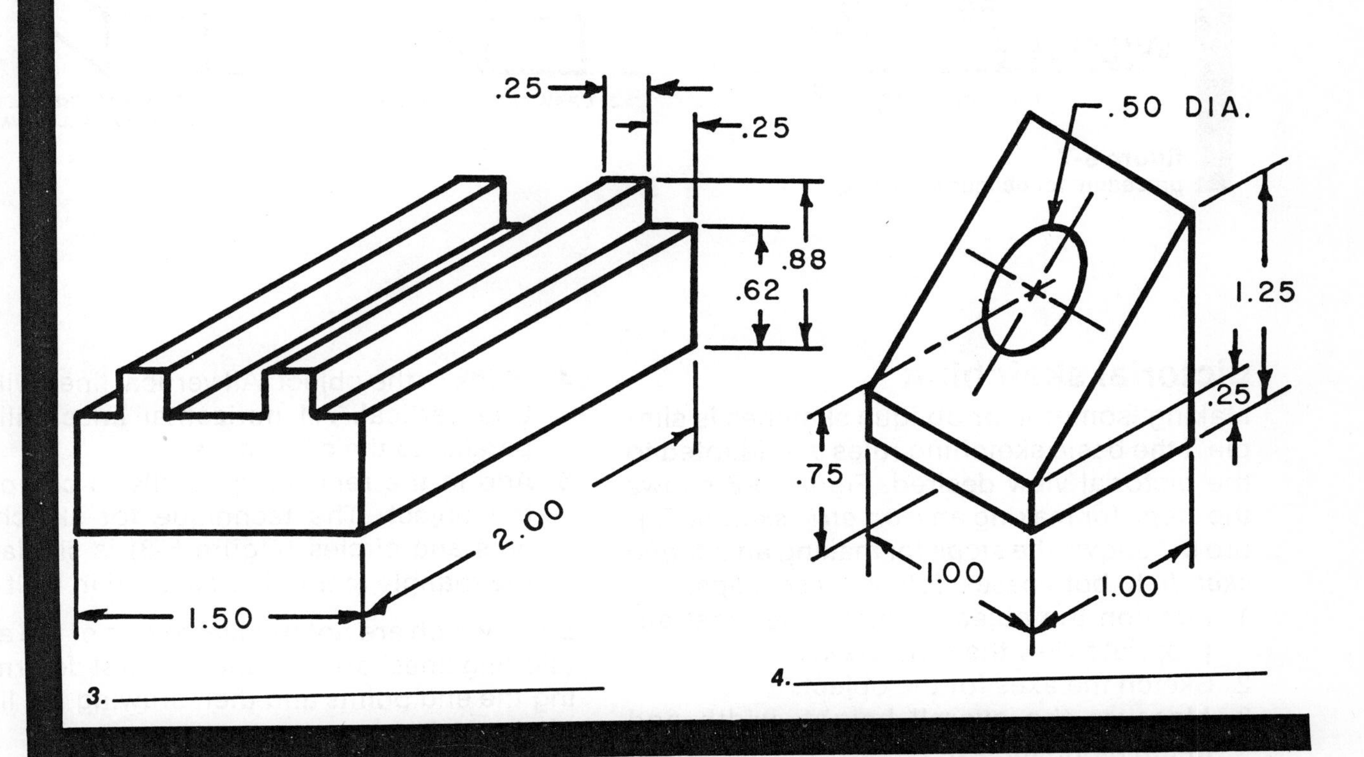

3. ______________________

4. ______________________

1. LAYOUT AXES.
2. MEASURE ALONG AXES.
3. COMPLETE THE SKETCH USING PARALLEL LINES.

figure 6-6
procedure for isometric sketching

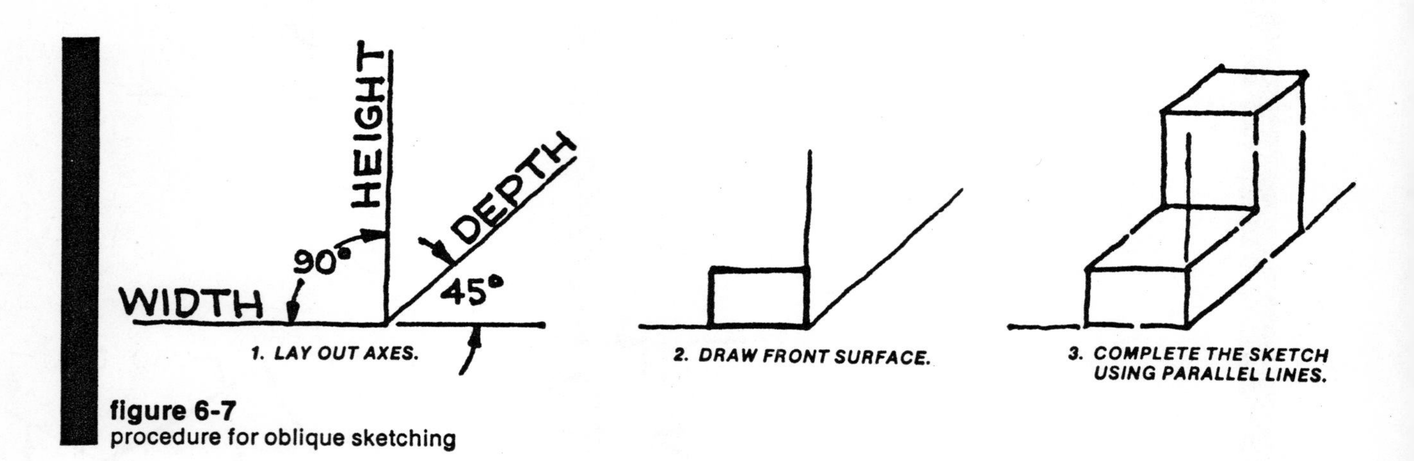

figure 6-7
procedure for oblique sketching

pictorial sketching

Making isometric or oblique sketches is simple if the basic sketching rules are adapted to the pictorial view desired. Figure 6-6 shows the steps for making an *isometric sketch*. Figure 6-7 shows the steps for making an *oblique sketch*. In both cases, follow these steps:

1. Position the object to make the most appropriate view the front view.
2. Sketch the axes for the object.
3. Measure the overall height, width and depth along the axes.
4. Block in the object. All vertical lines will be true vertical. All horizontal lines will be parallel to the other axes.
5. Add in the remaining details to complete the object. The technique for sketching arcs and circles (Figure 6-8) is similar to the triangle method (discussed in Unit 5).

Lines which are not parallel to one of the axes (sloping lines) are sketched by first determining the end points and then drawing the lines (Figure 6-9).

ISOMETRIC

OBLIQUE

figure 6-8
the triangle technique applied to pictorial sketching

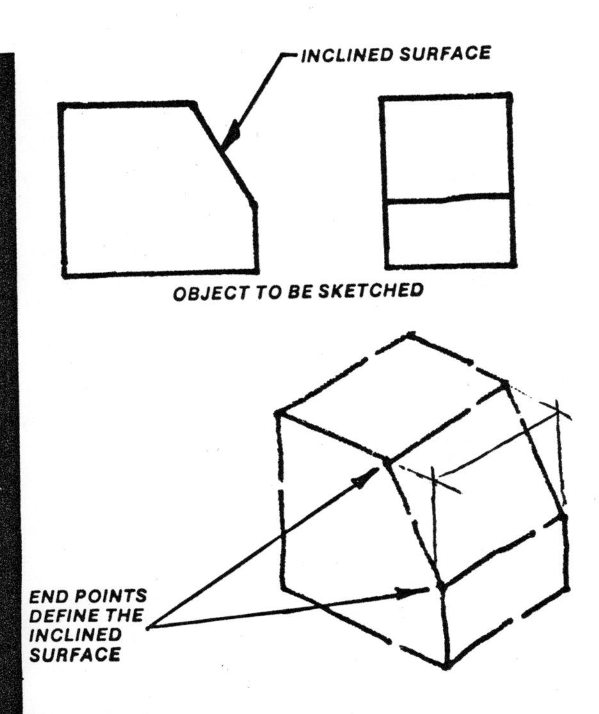

figure 6-9
sketching sloping surfaces by determining end points

exploded views

An easy-to-read, helpful view which is often used for assembly drawings is the *exploded view.* In this kind of view, the parts are shown

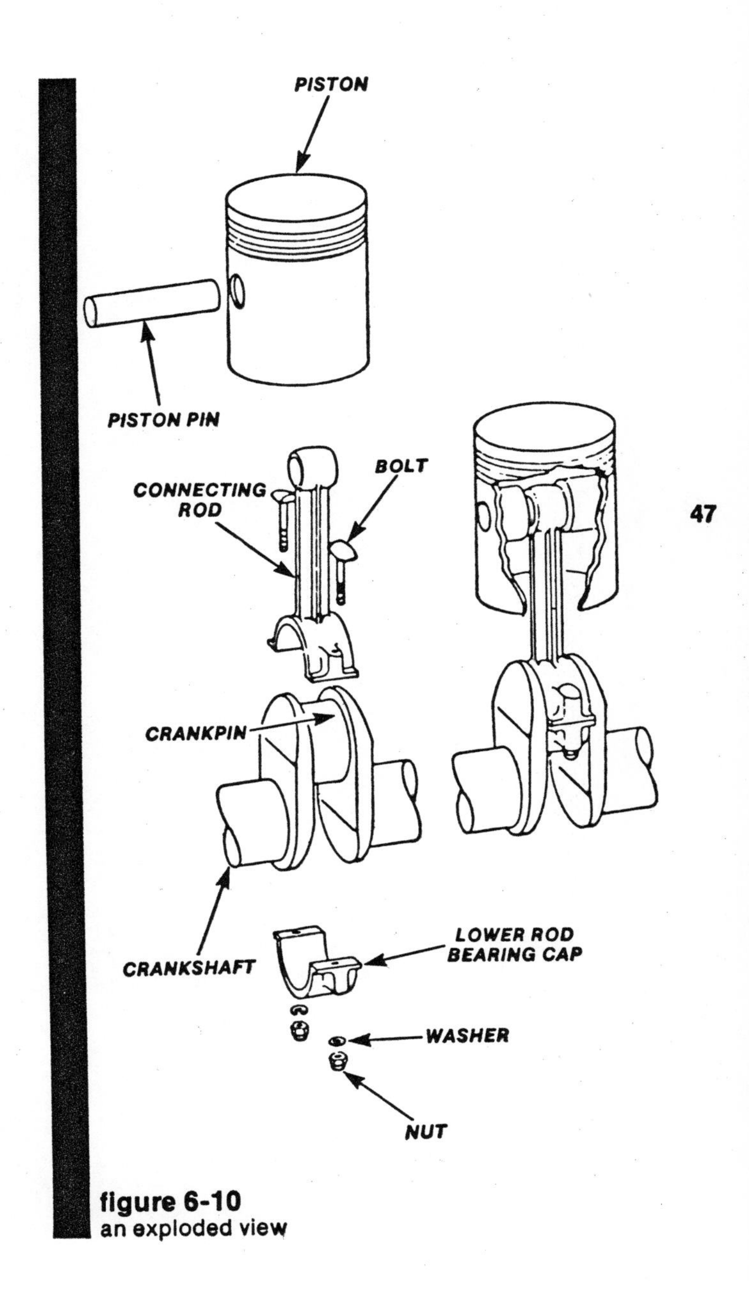

figure 6-10
an exploded view

pictorially, with all parts shown in their position relative to assembly (Figure 6-10). Exploded views are used most often in instruction books and service manuals.

This page for student notes

exercise 6-1

______________________________ name

Make isometric sketches of objects 2 and 3 of self-check quiz 6-a in the space below. Use the corner provided to start your sketch.

exercise 6-2

________________________ name

Make cavalier oblique sketches of objects 1 and 4 of self-check quiz 6-a. Start your sketch at the corner provided.

exercise 6-3

____________________________ name

Make cabinet oblique sketches of the objects shown below. Sketch the objects twice as large as shown, and maintain good proportion. Use a separate sheet of paper.

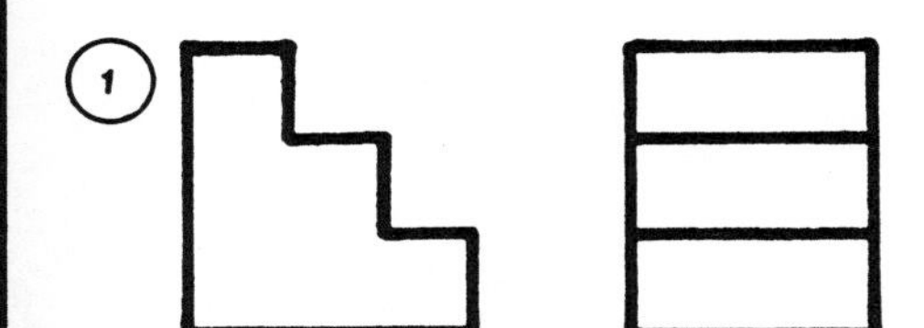

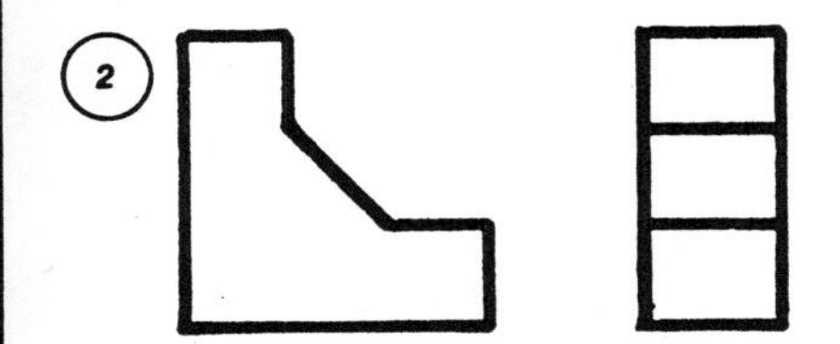

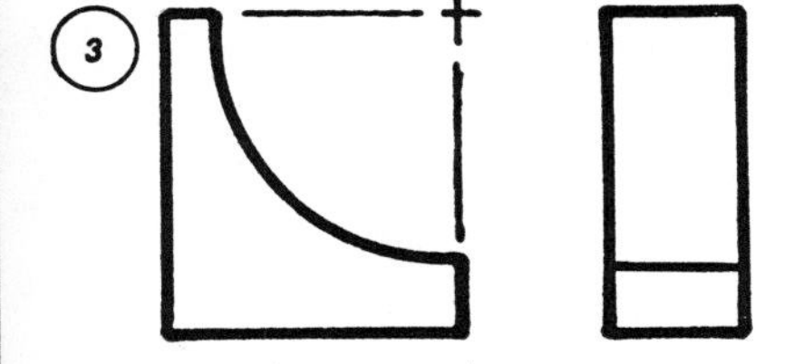

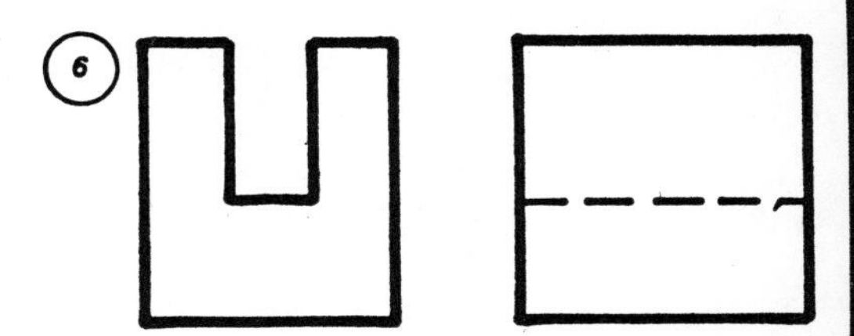

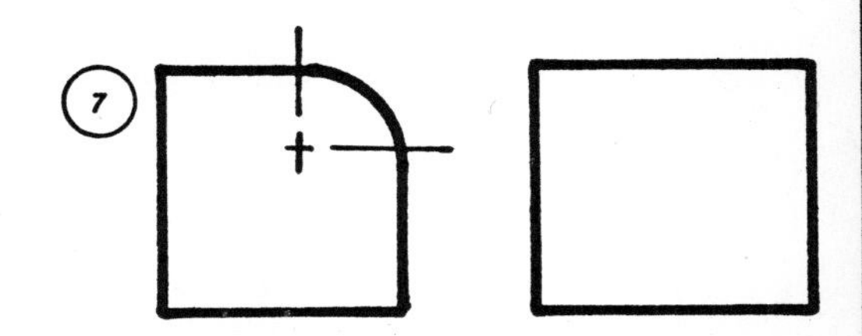

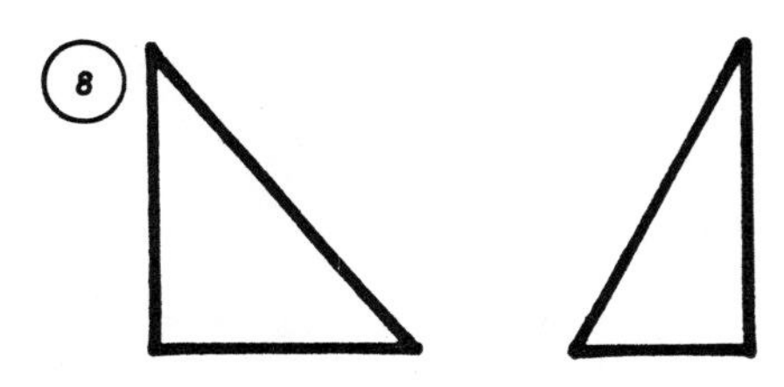

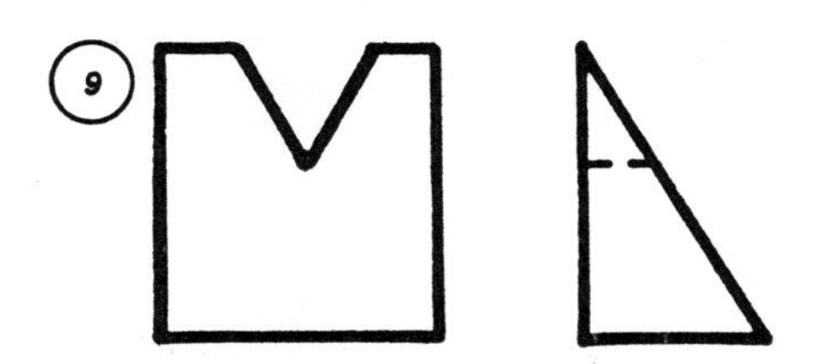

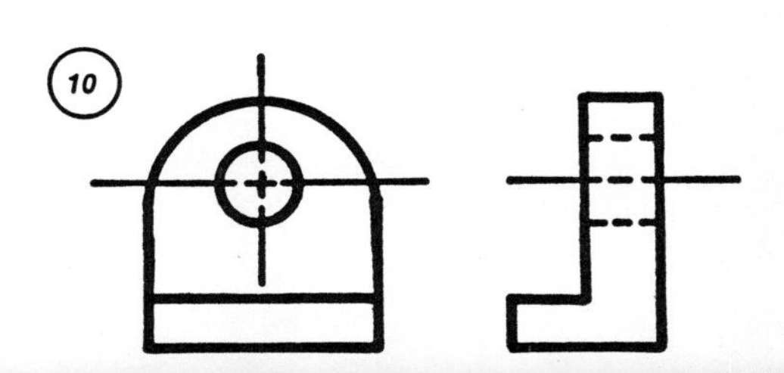

This page for student notes

drawing scales

Unit 7

Some objects are shown full size on drawings. Other objects too big for the drawing sheet are drawn to a reduced scale. Still other objects, such as wristwatch parts, require enlarged views. The *scale* of a drawing is the ratio of the size of the object on the drawing to the actual size of the object.

In this unit you will learn the application and interpretation of different scales used to make drawings: (a) architect's scale, (b) engineer's scale, (c) mechanical engineer's scale, and (d) metric scale. You will also learn to read the steel rule used by machinists in the shop.

the steel rule

Often a machinist must measure with a steel rule (Figure 7-1A). Some steel rules are graduated in 32nds and 64ths of an inch. Other steel rules are decimally divided. Some steel rules have combined metric/customary (inch) scales.

As a machinist, you will often use a steel rule to

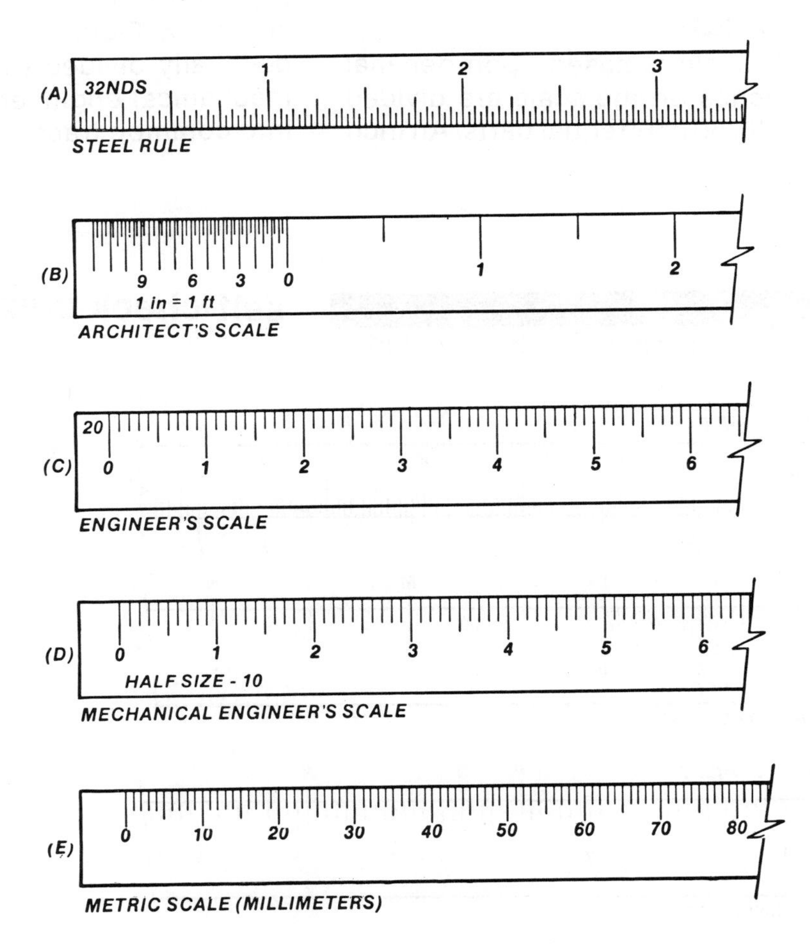

figure 7-1
different scales used to make drawings. The steel rule is for shop use.

measure a part but never to measure a drawing. If a dimension on a drawing is missing, advise the designer before proceeding with the machining operation.

architect's scale

The architect's scale is used primarily to draw building plans and details. With architect's scales, inches or fractions of an inch on the drawing represent one foot (12 inches) of the actual structure. Views of a building are usually drawn to a scale such as 1/8" = 1'-0" or 1/4" = 1'-0". Figure 7-1B shows an architect's scale. Here, 1 inch on the scale equals 1 foot on the building. Some machine shop drawings are drawn to architect's scales.

engineer's scale

The engineer's scale is based upon decimal dimensions. Inches on the scale are divided into 10, 20, 30, 40, 50 or 60 equal parts. An inch on one of these scales can represent inches, feet, miles, and so on. Figure 7-1C shows an engineer's scale. Here 1 inch on the scale equals 20 units on the actual object. These scales might appear on drawings as 1" = 1" (full scale), 1" = 30 miles or 1" = 500".

mechanical engineer's scale

The mechanical engineer's scale (sometimes called the mechanical drafter's scale) is often used for machine drawings. This scale is commonly graduated for full size, half size, quarter size, and eighth size (Figure 7-2).

Figure 7-1D shows a half-size mechanical engineer's scale. Note that on these scales, subdivisions represent fractions of an inch whereas on the architect's scale, subdivisions represent fractions of a foot. The scale shown in Figure 7-1D has the subdivisions decimally divided (tenths of an inch). Some mechanical engineer's scales are subdivided into common fractions (1/2, 1/4, 1/8, 1/16).

self-check quiz 7-a

Record the measurements indicated.

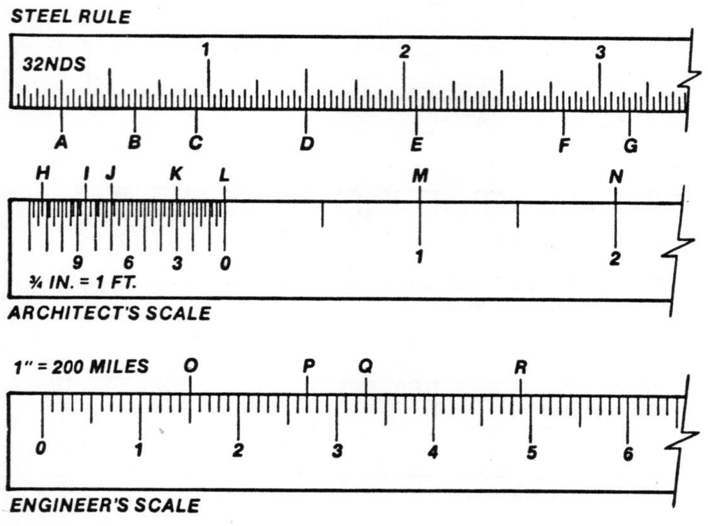

STEEL RULE

32NDS

1 2 3

A B C D E F G

H I J K L M N

9 6 3 0 1 2

¾ IN. = 1 FT.

ARCHITECT'S SCALE

1" = 200 MILES O P Q R

0 1 2 3 4 5 6

ENGINEER'S SCALE

A= ________
B= ________
C= ________
D= ________
E= ________
F= ________
G= ________
LM= ________
KM= ________
IM= ________
JL= ________
HN= ________
KL= ________
IK= ________
O= ________
P= ________
Q= ________
R= ________

FULL SIZE	**OR**	*1.00 = 1.00*	**OR**	*1 = 1*	**OR**	*1/1*	**OR**	*1:1*
HALF SIZE	**OR**	*.50 = 1.00*	**OR**	*1/2 = 1*	**OR**	*1/2*	**OR**	*1:2*
QUARTER SIZE	**OR**	*.25 = 1.00*	**OR**	*1/4 = 1*	**OR**	*1/4*	**OR**	*1:4*
EIGHTH SIZE	**OR**	*.125 = 1.00*	**OR**	*1/8 = 1*	**OR**	*1/8*	**OR**	*1:8*
TWICE SIZE	**OR**	*2.00 = 1.00*	**OR**	*2 = 1*	**OR**	*2/1*	**OR**	*2:1*
TEN TIMES SIZE	**OR**	*10.00 = 1.00*	**OR**	*10 = 1*	**OR**	*10/1*	**OR**	*10:1*

figure 7-2
specification of scales for machine shop drawings

metric scale

The metric scale is similar to the engineer's scale since the metric system is a decimal system. The basic unit of the metric system is the meter (39.37 inches). The metric scale is graduated in millimeters (1/1000 of a meter). Figure 7-1E shows a metric scale divided into millimeters (mm).

In Unit 17, you will learn more about the metric system and how it is used on drawings. Figure 7-3 shows a comparison of metric and customary measurements.

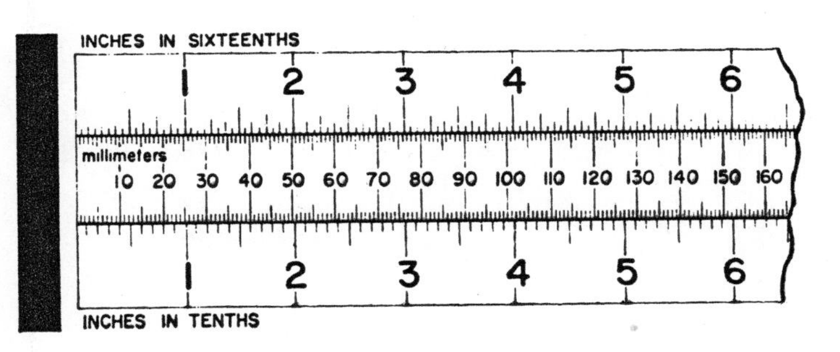

figure 7-3
a comparison of metric and customary measurements (1 inch equals 25.4 millimeters)

self-check quiz 7-b

Refer to Figure 7-3 and determine the closest corresponding inch measurements in tenths and in sixteenths for the following metric values:

	Inch (10ths)	Inch (16ths)
1. 20 millimeters		
2. 47 millimeters		
3. 71 millimeters		
4. 95 millimeters		
5. 127 millimeters		

This page for student notes

exercise 7-1

______________________________ name

Record the measurements indicated.

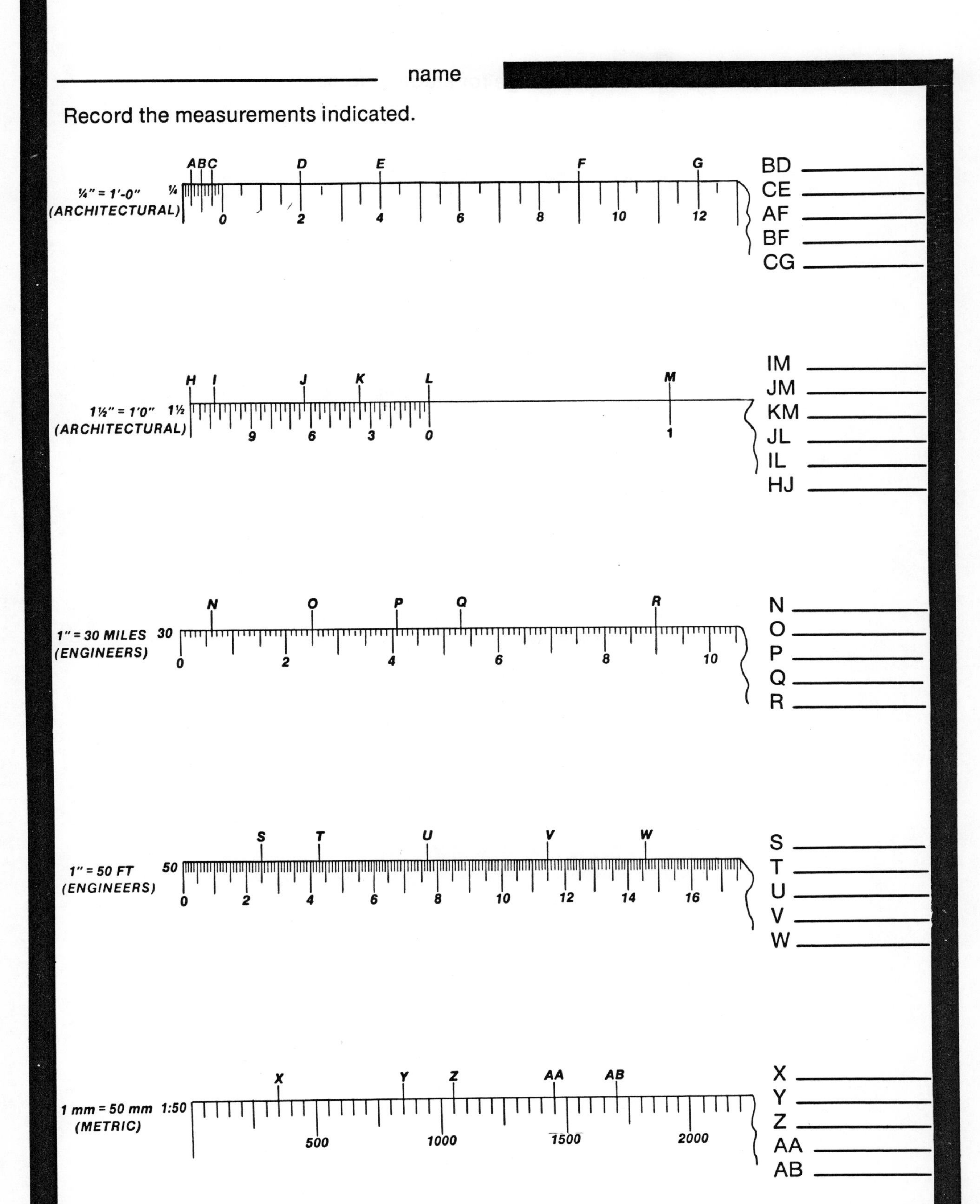

This page for student notes

math review

In order to read prints correctly, you must grasp certain math principles. In this unit you will review some of the basic math operations.

Test Your Skill. Work through self-check quiz 8a. If you have difficulty with any of the problems, review the material in the following pages.

self-check quiz 8-a

WHOLE NUMBERS

1. Add	555 + 323 + 103 + 878	______
2. Add	49 + 67792 + 5182 + 3	______
3. Subtract	581 from 882	______
4. Subtract	937 from 8421	______
5. Multiply	687 by 12	______
6 Multiply	5284 by 731	______
7. Divide	3000 by 148 (to nearest whole number)	______
8. Divide	176402 by 368 (to nearest whole number)	______

FRACTIONS

9. Add	3/16 + 5/16 + 7/8	______
10. Add	7/8 + 9/16 + 1/4	______
11. Add	8 3/16 + 10 1/8 + 5 1/2 + 2	______
12. Subtract	2 7/16 from 3 3/16	______
13. Subtract	7 7/8 from 11 1/4	______
14. Multiply	3/4 × 5/8	______
15. Multiply	2 3/16 × 1 1/4	______
16. Multiply	1 1/2 × 1 1/2 × 1 1/2	______
17. Divide	25 ÷ 1/4	______
18. Divide	1/4 by 25	______
19. Divide	3 1/8 by 1/16	______

DECIMALS

20. Add	3.60 + 12.45 + 5.65 + .45	______
21. Add	322.07 + 64.1 + 9.827 + 88	______
22. Subtract	832.17 from 973.78	______
23. Subtract	4762.016 from 6529.76	______
24. Multiply	1251 × 1.4	______
25. Multiply	65.7 × .25	______
26. Multiply	321.1 × 51.2	______
27. Divide	261.63 by 27	______
28. Divide	93.753 by 3.475	______
29. Divide	364.7 by 8.25 (3 decimal places)	______
30. Divide	.331 by .9	______

addition

For simple addition, line up the numbers in columns so that the units, tens, hundreds, and all like units are directly in line:

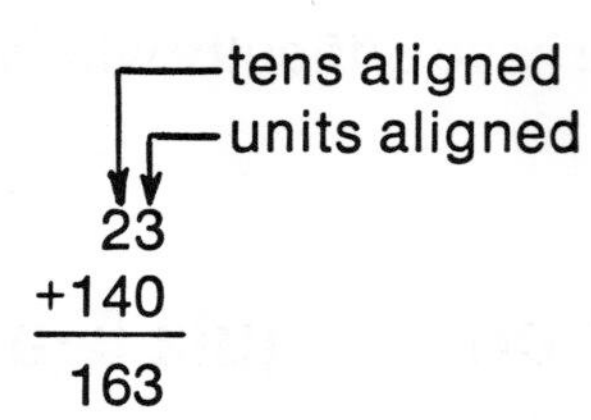

Begin adding each column starting at the right. When a column adds up to more than nine, the number of tens is carried to the column at the left. Thus in the following problem, 4 + 8 + 3 = 15. The 5 is recorded and the 1 (one ten) is carried to the tens column:

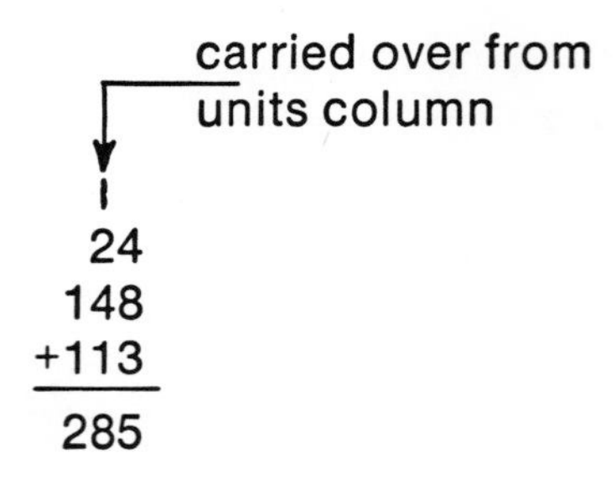

If the tens column adds up to more than nine, the hundreds are carried to the next left column, and so on.

When adding decimals, it is important to line up the decimal points. This ensures that the units, tens and hundreds are also lined up:

```
 1  1
205.6
  33.258
+ 12.06
250.918
```

subtraction

When subtracting, line up the numbers the same as you do for adding. Begin with the right column, subtracting the bottom number from the top one. If the top number is smaller, you will have to borrow from the column at the left:

```
  4
 953  (5 crossed out)
- 47
 906
```

In this example, 7 cannot be subtracted from 3. Therefore you must borrow 1 (one ten) from the tens column in order to subtract 7 from 13. When you subtract the tens column, remember that there is one ten less. In the example above, the 5 becomes 4 (four tens). Continue subtracting all columns in the same way.

You can check your subtraction by adding the answer to the bottom number. The result should be the top number. To check the problem above:

```
  1
 906
+ 47
 953
```

When subtracting decimals, again be sure the decimal points are aligned. If necessary, zeros can be added to the right of the decimal point without changing the value of the number. This may make it easier to subtract:

```
75.17 - 12.852 = 75.170
                -12.852
                 62.318
```

self-check quiz 8-b

1. 4.625 + 4.7
2. 49.906 + 18.56
3. 750 + 7.50 + .750
4. 228.5 - 114.25
5. Determine Dimension A

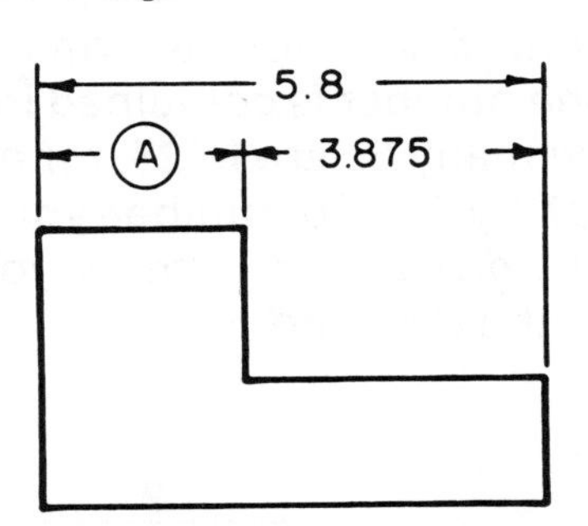

multiplication

When multiplying, begin with the bottom right-hand number (units). Multiply each of the top numbers: units, tens, hundreds, etc. Always place the first number of each product directly below the number by which you are multiplying:

92	
×514	
368	8 lines up with the 4
92	2 lines up with the 1
460	0 lines up with the 5
47288	

In this example, when multiplying by the 4, the first part of the product (4 × 2) lines up under the 4. When multiplying by the 1, the first part of the product lines up under the 1.

If you multiply decimals, you will have to determine where to place the decimal point in the answer. This is done by simply counting the number of decimal places to the right of the decimal in all the numbers multiplied:

6.258	3 decimal places
× .42	2 decimal places
12516	
25032	
2.62836	5 decimal places

In this example, three decimal places and two decimal places in the numbers multiplied, add up to five places in the answer.

self-check quiz 8-c

Determine these products:

1.) 27 × 4

2.) 600 × 20

3.) 43 × 417

4.) 5.25 × .18

5.) 3.75 × 2.1

6.) THE NOTCHES AND TEETH OF THE GAGE SHOWN BELOW ARE ALL THE SAME SIZE. DETERMINE THE OVERALL LENGTH L.

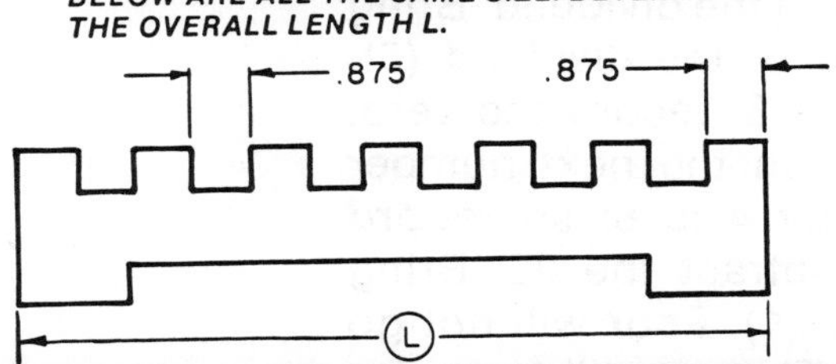

division

When you divide, you determine how many times one number is contained in another. To find how many 3s are in 27, for example, you divide 27 by 3. The number you divide by is called the *divisor*; the number you divide into is called the *dividend*.

```
              9
DIVISOR→3 ) 27←DIVIDEND
            27
             0
```

You know from the multiplication tables that 3 goes into 27 exactly 9 times. To check, multiply 3 times 9.

Sometimes, however, the answer doesn't come out to an exact number:

```
     1080¼ ← Remainder
4 ) 4321     Divisor
    4
     3
     0
     32
     32
       1
       0
       1 ← Remainder
```

To divide begin from the left of the dividend. In this example, the divisor 4 goes into the dividend 4 once. Record the 1. Multiply 1 × 4 and subtract from the 4 in the dividend. Bring down the next number of the dividend (3). Since 4 will not go into 3, record the zero, subtract, and bring down the next number (2). Now 32 divided by 4 is 8; so record the 8, multiply and subtract the 32. Bring down the final number (1). Four will not go into 1. Record the zero. There is still 1 remaining. This number is called a *remainder*. It can be expressed by a fraction. Place the remainder (1) over the divisor (4). This is the final fraction (¼).

If the divisor is a decimal number, you must first move the decimal point completely to the right to make it a whole number.

```
2.72, ) 81.05,6
```

Count the number of places you move the decimal point. Then move the decimal point an equal number of places in the dividend. In the example above the decimal point is moved two places. You may have to add zeros to make the required number of places:

```
3.865, ) 245.320,
```

The decimal point in the answer is placed directly above the relocated decimal point:

```
          29.8
2.72 ) 81.05 6
       544
       2665
       2448
        2176
        2176
           0
```

When there is a remainder in a division problem involving decimal numbers, you may continue adding zeros to the dividend until there is no remainder or until you have enough places in the answer:

```
        5.58
3.5 ) 19.530
      17 5
       203
       175
        280
        280
          0
```

self-check quiz 8d

Divide the following:

1.) $4\overline{)2032}$

2.) $7\overline{)7513}$

3.) $17\overline{)40}$

4.) $2.8\overline{)20.72}$

5.) $3.06\overline{)41.31}$

6.) *THE HOLES SHOWN BELOW ARE EQUALLY SPACED. DETERMINE DIMENSION A.*

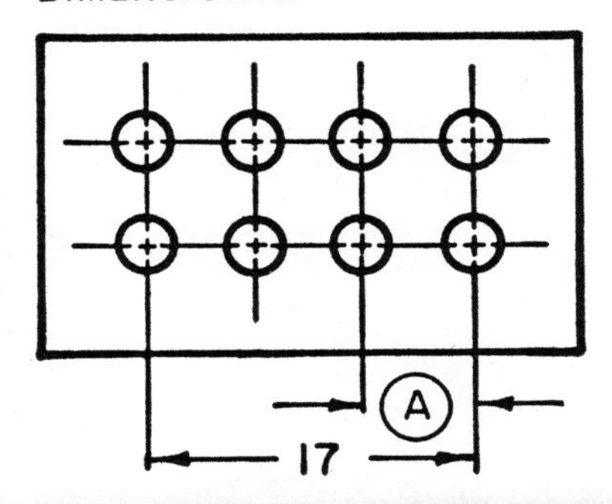

common fractions and decimal equivalents

Fractions are made of two numbers separated by a horizontal line. The top number is called the *numerator,* and the bottom number is the *denominator.* Common fractions can be converted to decimal fractions by dividing the numerator by the denominator:

$$\tfrac{1}{2} = 1 \div 2 = \quad 2\overline{)1.0} = .5 \text{ or } .50$$

(quotient .5; 10; 0)

$$\tfrac{3}{4} = 3 \div 4 = \quad 4\overline{)3.00} = .75$$

(quotient .75; 28; 20; 20; 0)

Decimal fractions can also be changed to common fractions. Set up the decimal in fractional form and then reduce it to the simplest fraction possible:

.25 = 25 hundredths = 25/100 = ¼
.875 = 875 thousandths = 875/1000 = ⅞

The following decimal equivalent chart (Figure 8-1) will help you in making conversions from common fractions to decimal fractions and from decimal fractions to common fractions. Usually for machining, decimal numbers are expressed in thousandths: 1/2 = .5 = .500 = 500 thousandths.

Conversion Table
Fractions—Decimals—Millimeters

Fractions					Decimals		mm	Fractions					Decimals		mm
4ths	8ths	16ths	32nds	64ths	to 2 places	to 3 place		4ths	8ths	16ths	32nds	64ths	to 2 places	to 3 places	
				1/64	0.02	0.016	0.3969					33/64	0.52	0.516	13.0968
			1/32		0.03	0.031	0.7937				17/32		0.53	0.531	13.4937
				3/64	0.05	0.047	1.1906					35/64	0.55	0.547	13.8906
		1/16			0.06	0.062	1.5875			9/16			0.56	0.562	14.2875
				5/64	0.08	0.078	1.9844					37/64	0.58	0.578	14.6844
			3/32		0.09	0.094	2.3812				19/32		0.59	0.594	15.0812
				7/64	0.11	0.109	2.7781					39/64	0.61	0.609	15.4781
	1/8				0.12	0.125	3.1750		5/8				0.62	0.625	15.8750
				9/64	0.14	0.141	3.5719					41/64	0.64	0.641	16.2719
			5/32		0.16	0.156	3.9687				21/32		0.66	0.656	16.6687
				11/64	0.17	0.172	4.3656					43/64	0.67	0.672	17.0656
		3/16			0.19	0.188	4.7625			11/16			0.69	0.688	17.4625
				13/64	0.20	0.203	5.1594					45/64	0.70	0.703	17.8594
			7/32		0.22	0.219	5.5562				23/32		0.72	0.719	18.2562
				15/64	0.23	0.234	5.9531					47/64	0.73	0.734	18.6532
1/4					0.25	0.250	6.3500	3/4					0.75	0.750	19.0500
				17/64	0.27	0.266	6.7469					49/64	0.77	0.766	19.4469
			9/32		0.28	0.281	7.1437				25/32		0.78	0.781	19.8433
				19/64	0.30	0.297	7.5406					51/64	0.80	0.797	20.2402
		5/16			0.31	0.312	7.9375			13/16			0.81	0.812	20.6375
				21/64	0.33	0.328	8.3344					53/64	0.83	0.828	21.0344
			11/32		0.34	0.344	8.7312				27/32		0.84	0.844	21.4312
				23/64	0.36	0.359	9.1281					55/64	0.86	0.859	21.8281
	3/8				0.38	0.375	9.5250		7/8				0.88	0.875	22.2250
				25/64	0.39	0.391	9.9219					57/64	0.89	0.891	22.6219
			13/32		0.41	0.406	10.3187				29/32		0.91	0.906	23.0187
				27/64	0.42	0.422	10.7156					59/64	0.92	9.922	23.4156
		7/16			0.44	0.438	11.1125			15/16			0.94	0.938	23.8125
				29/64	0.45	0.453	11.5094					61/64	0.95	0.953	24.2094
			15/32		0.47	0.469	11.9062				31/32		0.94	0.969	24.6062
				31/64	0.48	0.484	12.3031					63/64	0.98	0.984	25.0031
1/2					0.50	0.500	12.7000	1					1.00	1.000	25.4000

figure 8-1
decimal equivalent chart

manipulating fractions

When you add or subtract common fractions, all the fractions must have the same denominator. Then the numerators are added (or subtracted, as the problem requires). Finally the answer is reduced to its lowest fractional form:

1.) $4/5 + 7/9 = 36/45 + 35/45$ Change to common denominator

$= \frac{36 + 35}{45}$ Add numerators

$= 71/45$

$= 1\ 26/45$ Reduce to lowest terms

2.) $9\ 5/8 + 2\ 1/4 = 77/8 + 9/4$

$= 77/8 + 18/8$

$= \frac{77+18}{8}$

$= 95/8$

$= 11\ 7/8$

3.) $10\ 2/3 - 4\ 1/9 = 32/3 - 37/9$

$= 96/9 - 37/9$

$= \frac{96-37}{9}$

$= 59/9$

$= 6\ 5/9$

self-check quiz 8-e

Perform the indicated operations:

1. 473.1 + 32.85 + 20 ______
2. 8001 - 98.45 ______
3. 2/3 + 1/2 ______
4. 4/5 + 7/8 + 1/16 ______
5. 5/8 + 1/4 - 1/32 ______

Change the following to decimal equivalents:

6. 1/1000 ______
7. 1/10 ______
8. 1/10000 ______
9. 1/100 ______
10. 1/5 ______
11. 1/20 ______
12. 3/10000 ______
13. 5/1000 ______
14. 6/10 ______
15. 12/10000 ______

rounding off

When a common fraction is converted to its decimal equivalent, the number often has more decimal places than needed. Example: 1/32 = .03125. In such a situation, the decimal equivalent is rounded off to a less precise value.

When the figure to the right of the last significant digit is *less than five,* the last significant digit remains unchanged. Example: 6.234375, if held to three significant digits, would 6.234.

When the figure to the right of the last significant digit is *more than five,* the last significant digit is increased by one. Example: 3.1875, if held to two significant digits, would be 3.19.

When the figure beyond the last significant digit is *exactly five,* the last significant digit is increased by one if it is odd, and it remains unchanged if it is even. Example: 7.625, if held to two significant digits, would be 7.62. If held to two significant digits, 4.375 would be 4.38.

additional examples:

.125 to the nearest hundredth = .12
2.218 to the nearest hundredth = 2.22
8.848 to the nearest tenth = 8.8
.5625 to the nearest thousandth = .562
4.984375 to the nearest ten-thousandth = 4.9844

Note: Always machine to the accuracy required by the tolerances shown on the drawing. Tolerances are discussed in Unit 11.

angular measurements

Circles, holes, curves, and other circular objects are measured in a 60-based system rather than a 10-based system. This will pose no problem to you since you are familiar with a similar 60-based system for telling time: 60 seconds = 1 minute; 60 minutes = 1 hour.

A complete circle contains 360 degrees (360°). Consider a circle as a pie. A fourth of the pie equals 90° (¼ of 360°). Likewise half of the pie equals 180° (½ of 360°). See Figure 8-2.

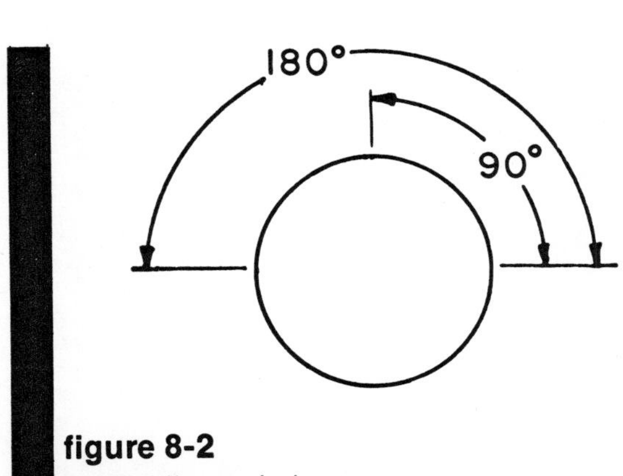

figure 8-2
subdividing a circle

Each degree can be subdivided into 60 equal parts called minutes (1° = 60′). Each minute can be subdivided into 60 equal parts called seconds (1′ = 60″).

Just as in our time system, addition and subtraction of circular dimensions requires expressing the dimensions in like terms and then performing the simple arithmetic. For example,

Subtract 40 degrees, 45 minutes, and 20 seconds from 180 degrees:

$$\begin{array}{r} 180^\circ \\ -\ 40^\circ\ 45'\ 20'' \\ \hline \end{array} = \begin{array}{r} 179^\circ\ 59'\ 60'' \\ -\ 40^\circ\ 45'\ 20'' \\ \hline 139^\circ\ 14'\ 40'' \end{array} \text{ (answer)}$$

exercise 8-1

__________________________ name

Change the following to mixed numerals (Reduce to lowest form):

______ 1) 17/4

______ 2) 11/8

______ 3) 16/4

Change the following to improper fractions (express totally as fractions):

______ 4) 1 1/2

______ 5) 4 1/4

Reduce the following to lowest fractional form:

______ 6) 15/60

______ 7) 16/24

______ 8) 16/32

Addition and Subtraction:

______ 9) 4.625 + 4.7

______ 10) 45.906 + 18.56

______ 11) 750 + 7.50 + .750

______ 12) 5.8 - 3.875

______ 13) 228.5 - 114.25

Round off the following to the nearest hundredth:

______ 14) .0156

______ 15) .03125

______ 16) .0625

Round off the following to the nearest thousandth:

______ 17) .21875

______ 18) .28125

______ 19) 7.609375

______ 20) 2.78125

Express the following as decimals:

______ 21) 29/64

______ 22) 7/32

______ 23) 19/32

Express the following as fractions:

______ 24) .1250

______ 25) .875

exercise 8-2

______________________ name

Perform the required operations.

1. 61 × 3 = ____________
2. 6.1 × 30.2 = ____________
3. .50 × .875 = ____________
4. .25 × 50 = ____________
5. 28 ÷ 10 = ____________
6. 45 ÷ 8.3 = ____________
7. Express 7/9 as a decimal ____
8. Express 4/3 as a decimal ____
9. Express .14 as a fraction ____
10. Express 1.506 as a mixed numeral ____________
11. 360 ° -45° = ____________
12. 10 ° -1°47′ = ____________
13. 45 ° -22°29′58″ = ____________

Convert the following to fractions.

14. .0625 ____________
15. .15625 ____________
16. .9219 ____________
17. .7031 ____________

Convert the following to decimals.

18. 3/64 ____________
19. 13/32 ____________
20. 15/16 ____________

Round off the following to the nearest thousandth.

21. .9375 ____________
22. .8125 ____________
23. 7.5625 ____________
24. 4.4375 ____________
25. 1.1975 ____________

exercise 8-3

______________________ name

DETERMINE THE DESIRED DIMENSIONS

A ________
B ________
C ________
D ________
E ________
F ________

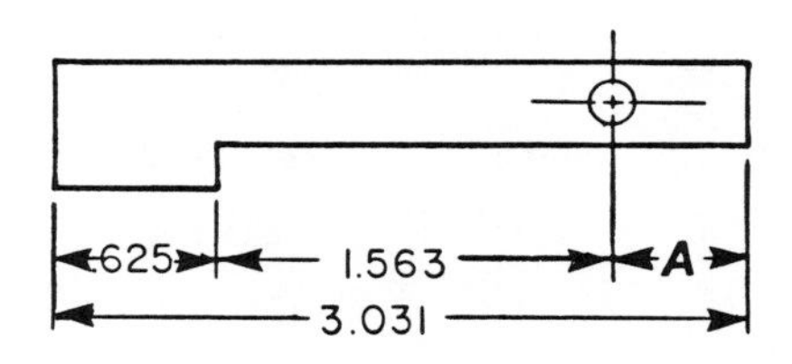

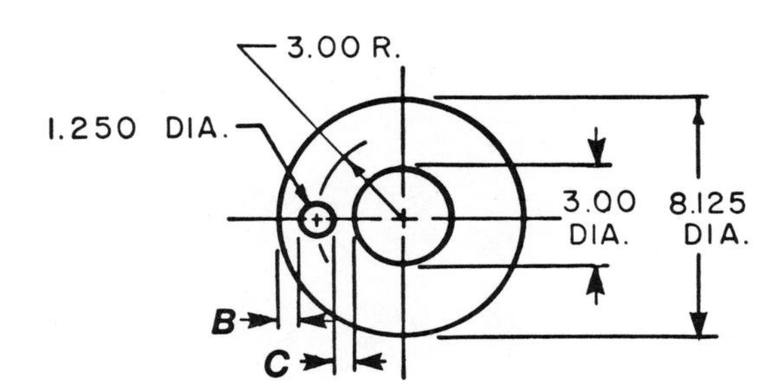

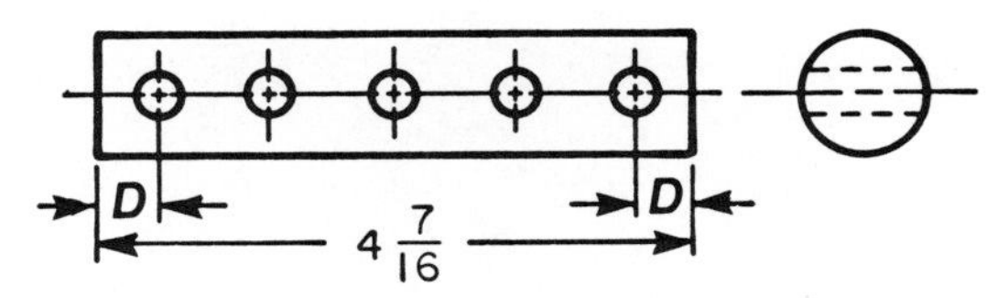

CENTER-TO-CENTER OF HOLES = $\frac{7}{8}$

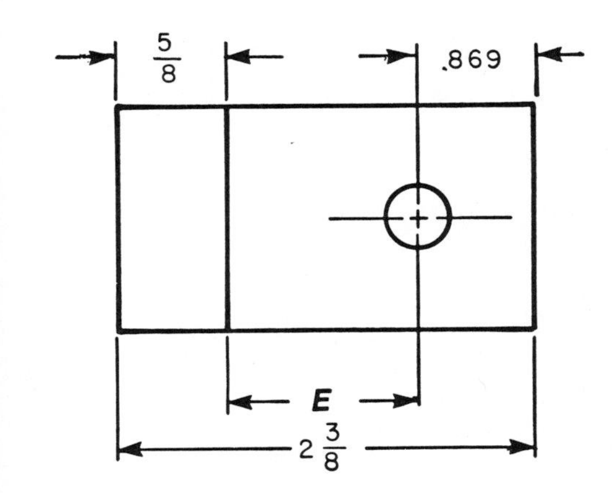

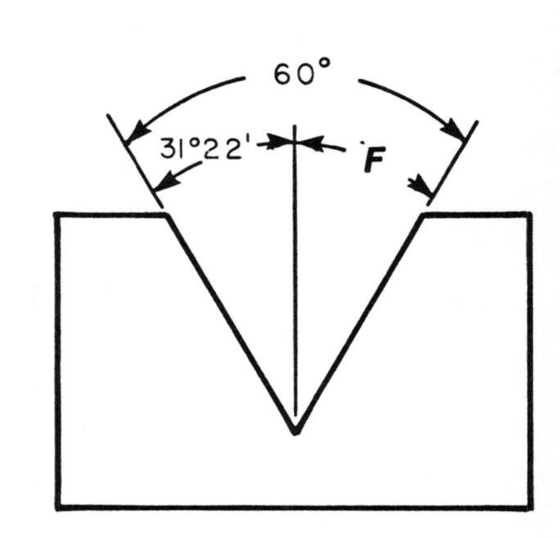

This page for student notes

Unit 9

dimensions

In order to machine a part, you must have a complete description of it. You get this description from views, notes and dimensions. Views give the shape description. Notes define processes, materials, and other shop information. Dimensions describe size and location. After completing this unit you will be able to interpret basic dimensioning practices.

size and location dimensions

Dimensions on a drawing provide either size information or location information. Size dimensions limit the various parts of the object. Location dimensions define the position of these parts in relation to one another. Figure 9-1 shows simple objects with size and location dimensions indicated.

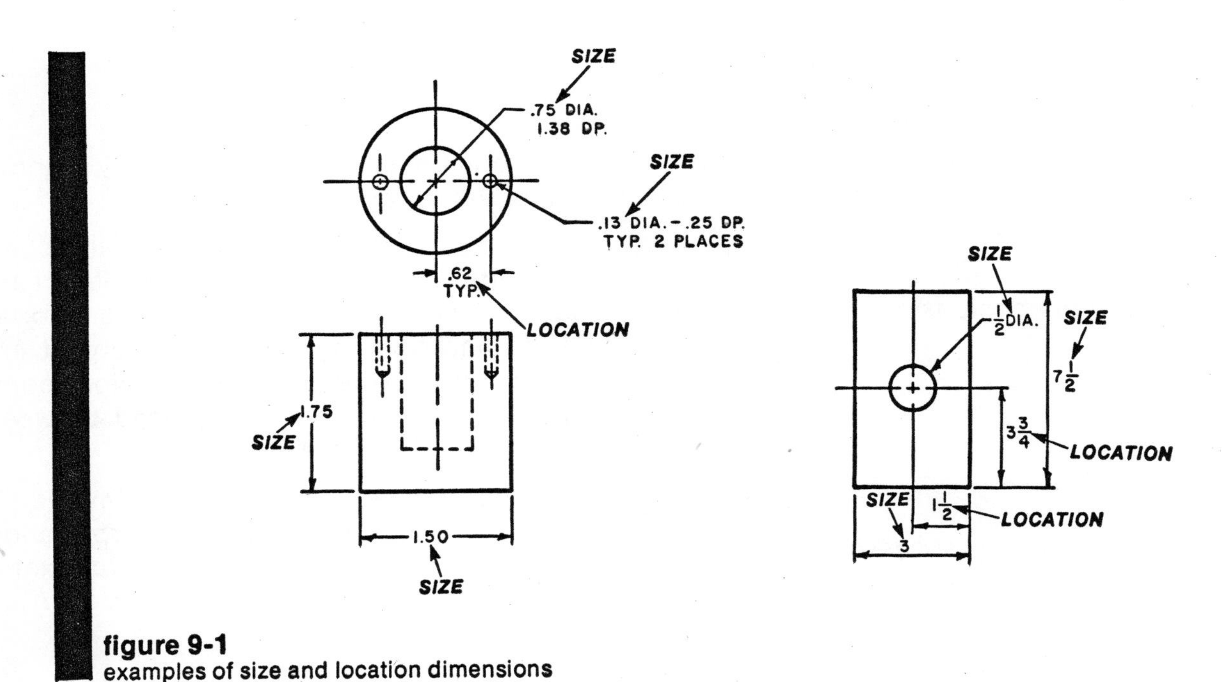

figure 9-1
examples of size and location dimensions

dimensioning practices

The drafter usually places dimensions outside the views so that the dimensions do not interfere with or confuse the drawing (Figure 9-2). Two different methods of dimensioning commonly used are aligned dimensioning and unidirectional dimensioning.

Aligned dimensions are placed parallel to the feature being measured. Horizontal dimen-

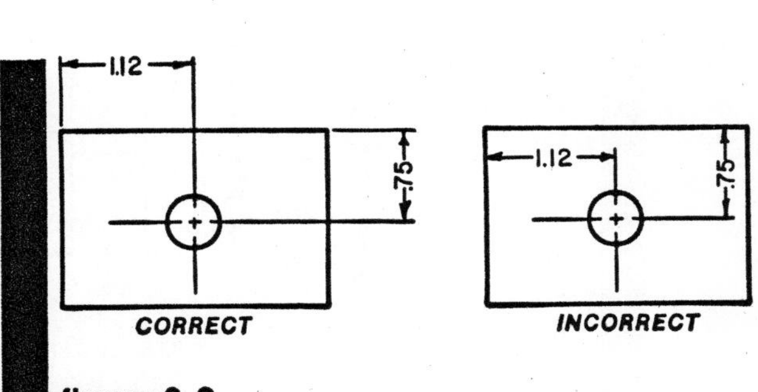

figure 9-2
dimensions are normally placed off the views to avoid confusion

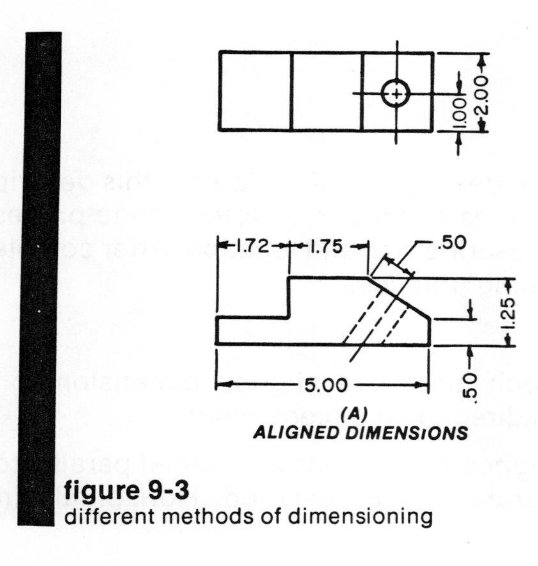

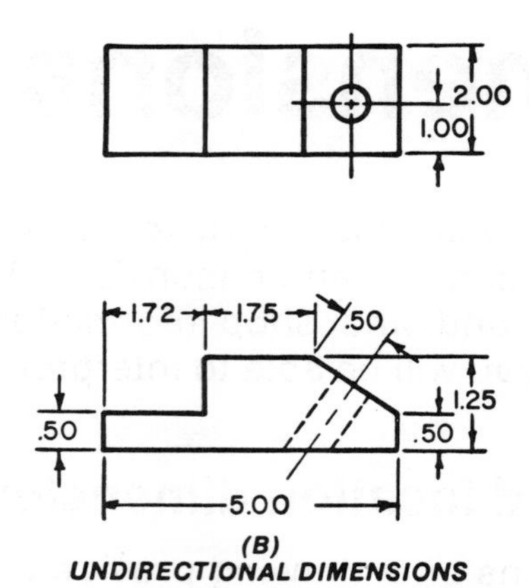

figure 9-3
different methods of dimensioning

sions are read horizontally from left to right. Vertical dimensions, however, appear sideways on the drawing and must be read from the right side (Figure 9-3A).

Unidirectional dimensions are placed so that all dimensions are read horizontally without turning the drawing (Figure 9-3B).

Overall dimensions which describe the complete width, height and depth of an object are usually placed farthest from the views. Shorter dimensions defining part of the object are placed closer to the object. Dimensions common to adjacent views are placed between those views. See Figure 9-4.

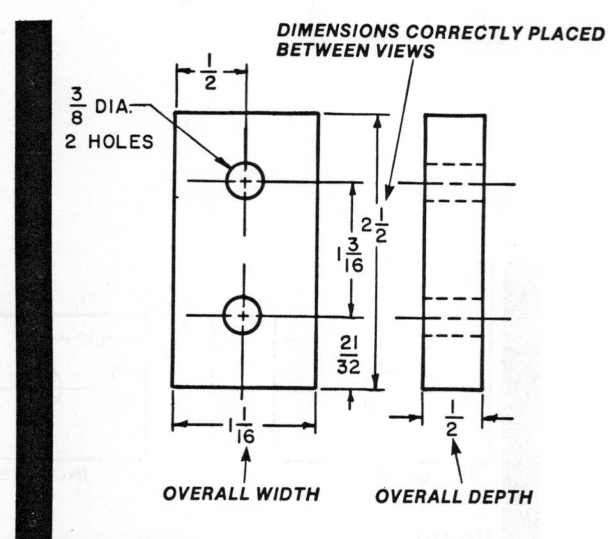

figure 9-4
correct placement of dimensions

It is common practice to omit inch marks (") except when confusion or misinterpretation could result. One-inch dimensions, for example, usually show inch marks (Figure 9-5A).

reference dimensions

Machine shop drawings usually show only those dimensions needed to machine the parts. Sometimes, however, the drafter gives non-machining dimensions which provide useful information to the machinist. Such dimensions have the designation REF, which means they are for reference only (Figure 9-5B) and are not for machining or inspection purposes.

Usually, the drafter gives a dimension in only one view but occasionally may repeat it in another view for the convenience of the mach-

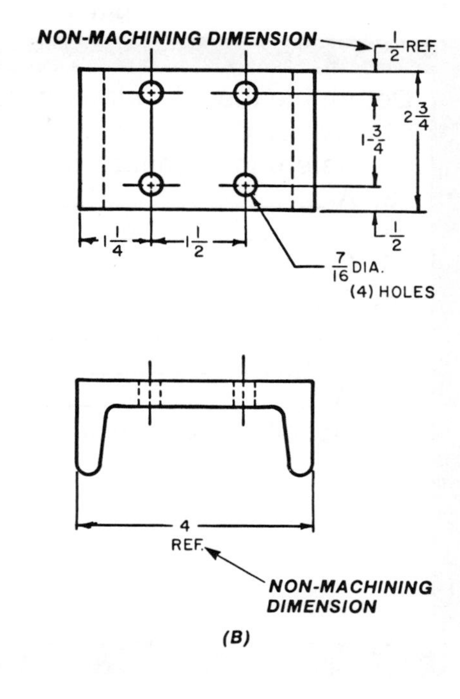

figure 9-5
representation of reference dimensions

inist. In this case, the second occurrence of the dimension is noted REF (Figure 9-5A).

typical dimensions

Sometimes a feature on a part is repeated. When this happens, the drafter dimensions the feature in one place and shows that the dimension applies in the other locations by adding the designation TYP, which stands for "typical." In Figure 9-6 (top) only one of the small holes is dimensioned. The other small hole is understood to have the same dimension because of the designation TYP.

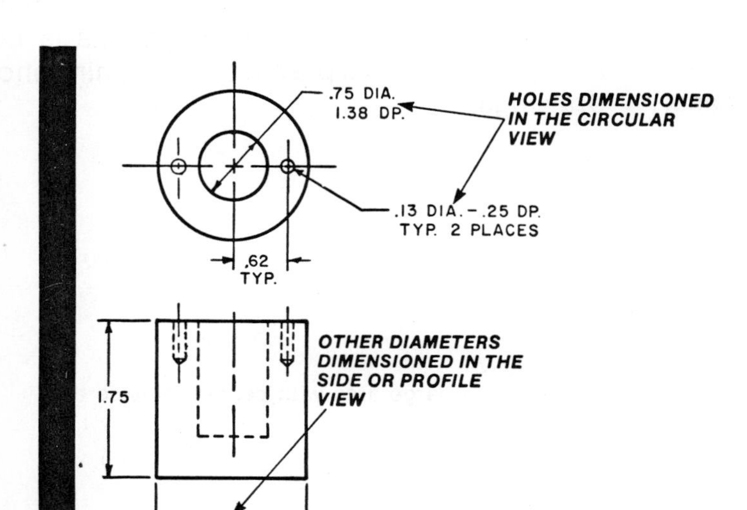

figure 9-6
specifying diameters

self-check quiz 9-a

Determine the following dimensions:

1. Overall dimensions
2. Reference dimension
3. A ____________________
4. B ____________________
5. C ____________________
6. D ____________________
7. E ____________________
8. F ____________________
9. G ____________________
10. List all size dimensions

dimensioning curved surfaces

The drafter dimensions arcs and curved surfaces in views which show the true shape best. The letter R or the abbreviation RAD is used to denote a radius. The abbreviation DIA, the symbol ϕ or the letter D stands for a diameter. Diameters are usually defined in their side or profile views except for holes. When it is obvious from the view that a dimension refers to a diameter, the term DIA is omitted. See Figure 9-6 (bottom).

Occasionally a curved surface has a radius which falls outside the drawing. In this case, the dimension line is broken or offset as shown in Figure 9-7.

dimensioning limited spaces

Normally a dimension is placed in the middle of the dimension line. However, space limitations often require other dimensioning techniques as shown in Figure 9-8. Some drafters place dimensions above or beside the dimension line (Figure 9-9), although this is not accepted industrial practice for machine shop drawings.

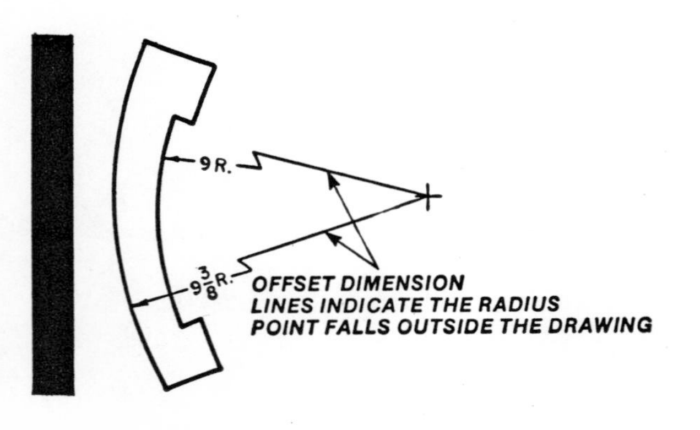

figure 9-7
dimensioning large arcs with centers outside the drawing

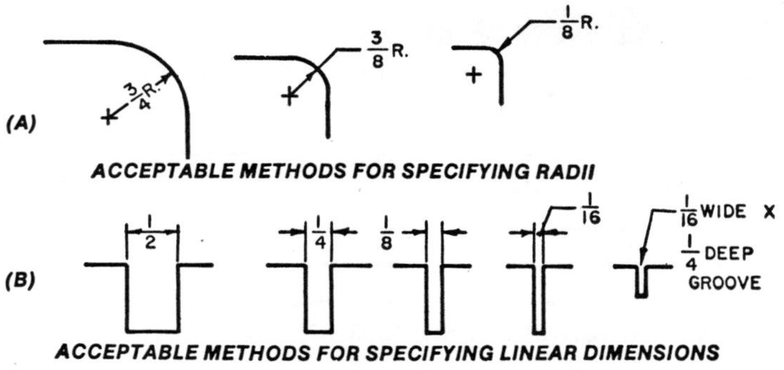

(A)
3/4 R.
3/8 R.
1/8 R.
ACCEPTABLE METHODS FOR SPECIFYING RADII
(B)
1/2
1/4
1/8
1/16
1/16 WIDE X
1/4 DEEP
GROOVE
ACCEPTABLE METHODS FOR SPECIFYING LINEAR DIMENSIONS

figure 9-8
dimensioning practices for confined areas

1/2 DIA.
7 1/2
3 3/4
1 1/2
3

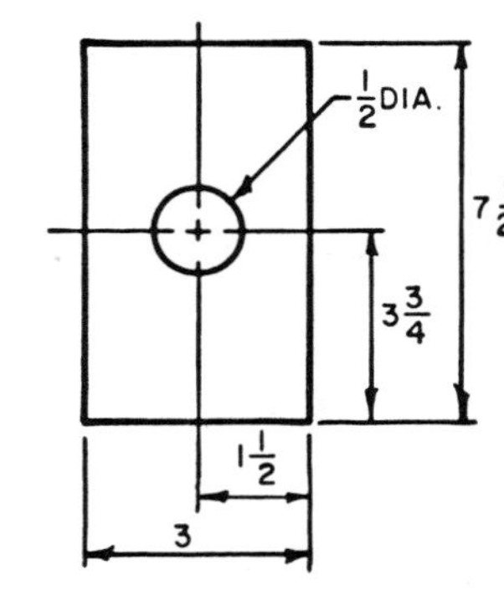

figure 9-9
dimensions placed above and beside the dimension line

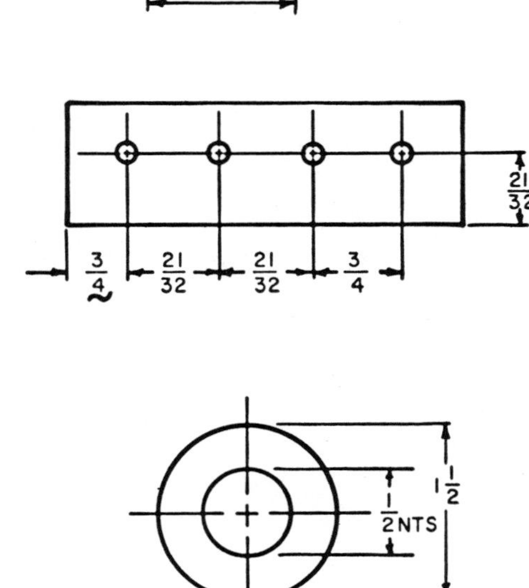

21/32
3/4
21/32
21/32
3/4
1 1/2
1/2 NTS

figure 9-10
different ways of noting a dimension which is not to scale

dimensions out-of-scale

A drafter may have to change a dimension after a drawing is completed. In such cases, it is not economical to redraw the drawing to correct the views. The drafter identifies those dimensions which are out of proportion by underlining the dimension(s) with a heavy wavy line or by adding the abbreviation NTS (Not To Scale) after the dimension(s) (Figure 9-10).

self-check quiz 9-b

1. Are the dimensions on this drawing aligned or unidirectional? __________
2. Determine the following dimensions.
   A __________
   B __________
   C __________
   D __________
   E __________
3. Are there any location dimensions on this drawing?
   ____
4. Give the overall dimensions. __________
5. Sketch the top view of the object.
6. Sketch the side view of the object.

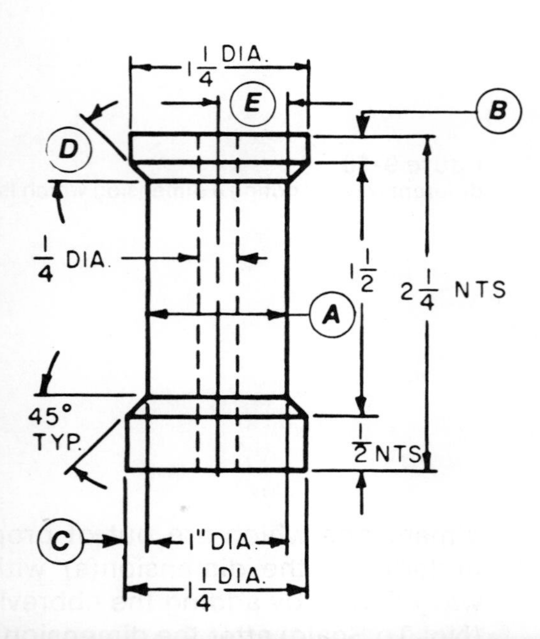

exercise 9-1

____________________________ name

1. Are the dimensions on this drawing aligned or unidirectional? __________
2. Name the views shown. ____
3. Identify each of the following lines as to type of line:

A __________
B __________
C __________
D __________
E __________

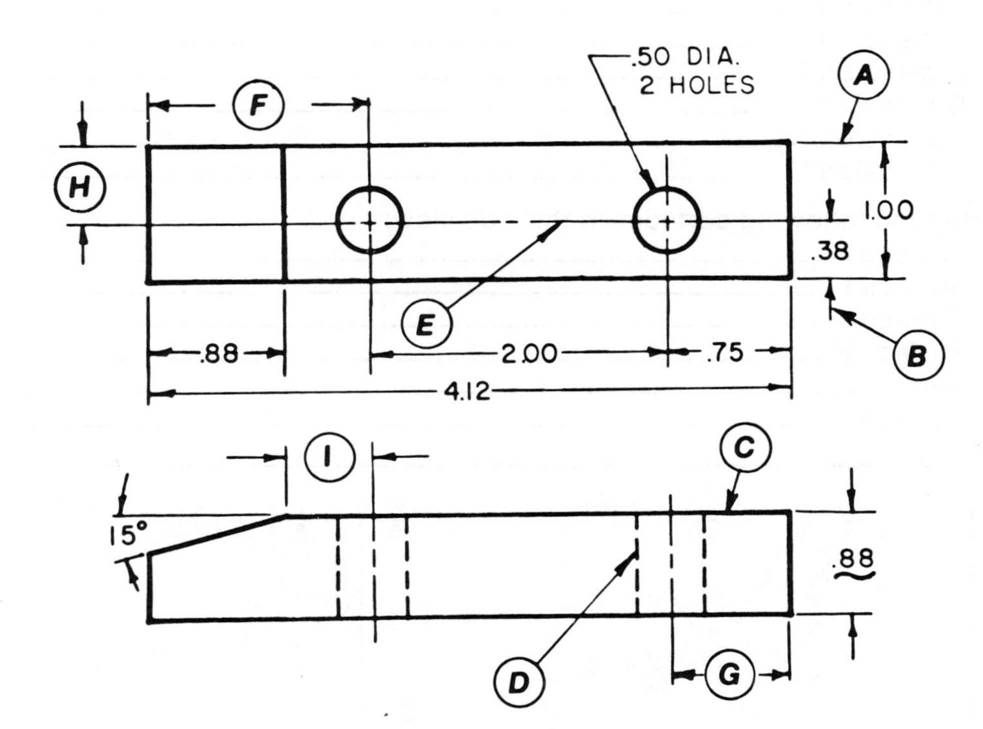

4. Determine the following dimensions:

F __________
G __________
H __________
I __________

5. List the overall dimensions. ____
6. Which dimensions are not to scale? __________________________________
7. How far apart are the .50 DIA holes? ________________________________
8. List all size dimensions given. ______________________________________
9. List all location dimensions given. ___________________________________

exercise 9-2

name

Identify the following:

1. Drawing Number
2. Name of the part
3. Number of parts required
4. Casting number
5. Original drawing number
6. Dimensions not to scale
7. Reference dimension
8. Dimension A
9. Dimension B
10. Dimension C
11. Dimension D
12. Dimension E
13. Dimension F
14. Dimension G

Identify the following surfaces in the side view:

15. Surface H
16. Surface I
17. Surface J
18. Surface K
19. Surface L
20. Surface M

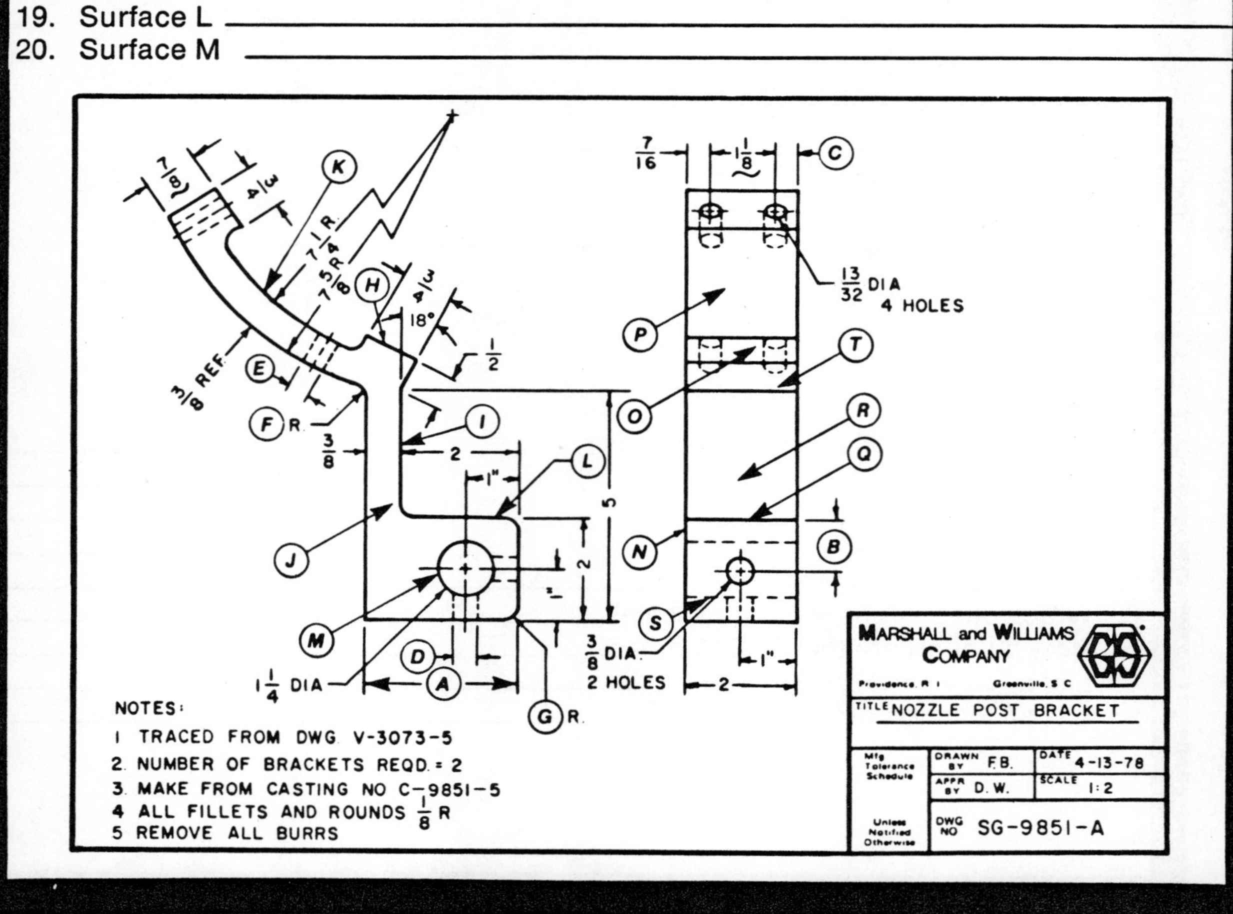

dimensioning holes

Unit 10

The machinist makes holes in several different ways such as drilling, reamining, boring, and punching. In this unit you will learn what some of these terms mean, and how they are dimensioned on a drawing. The drafter usually gives only the hole size required and leaves the shop process up to the machinist. Use of the terms *drill, ream* or *bore* on a drawing indicates the accuracy and smoothness required, since reaming and boring are shop processes for precise, smooth holes.

The notation for holes is specified in this order: (1) hole diameter, (2) shop process (if specified), (3) hole depth, if not a through-hole (designated "thru-hole"), (4) number of holes if more than one (Figure 10-1).

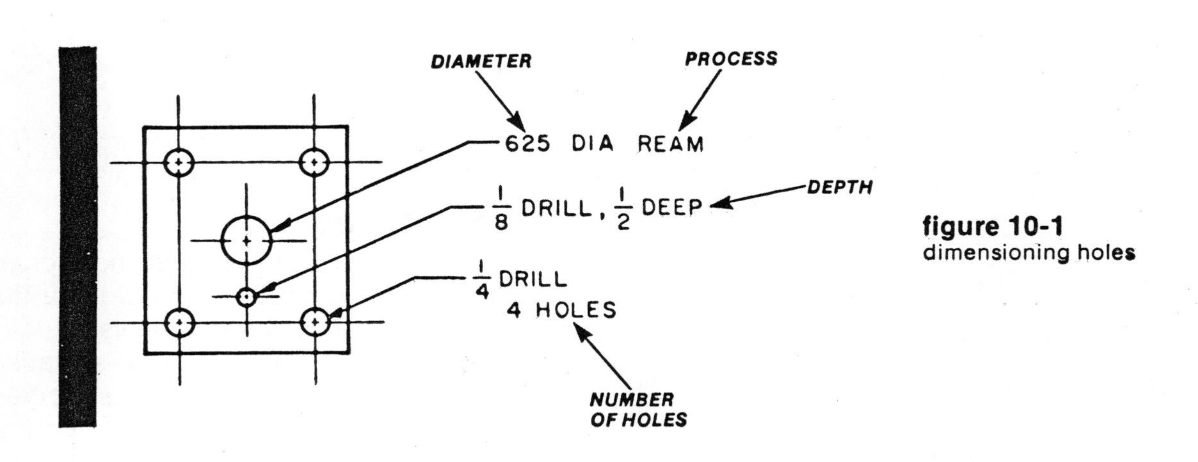

figure 10-1
dimensioning holes

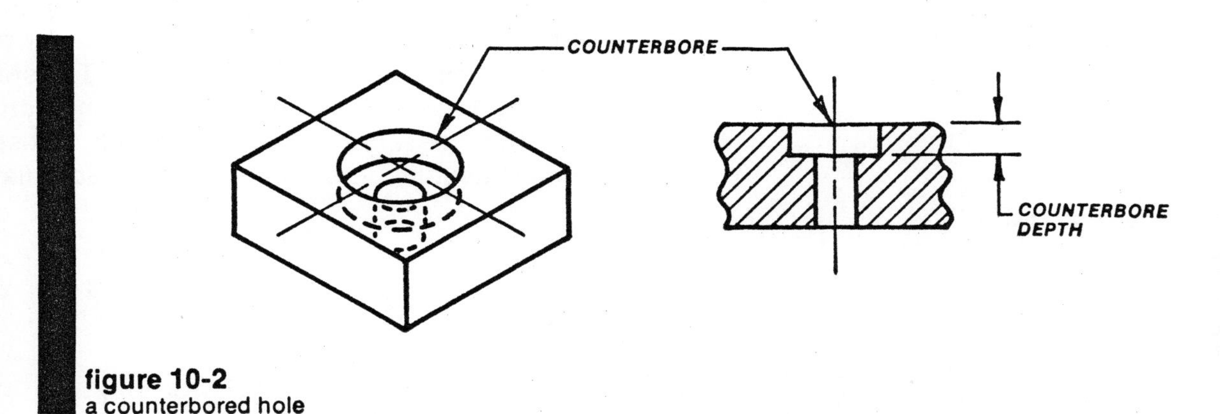

figure 10-2
a counterbored hole

counterbored holes

A counterbore (Figure 10-2) is a recess for a bolt-head, nut, or other fastener. The notation for a counterbored hole is specified in this order: (1) hole diameter (and depth if not a through-hole), (2) diameter of counterbore, (3) depth of counterbore, (4) number of holes if more than one (Figure 10-3). Note that the leader comes to the edge of the counterbore, not to the edge of the drilled hole.

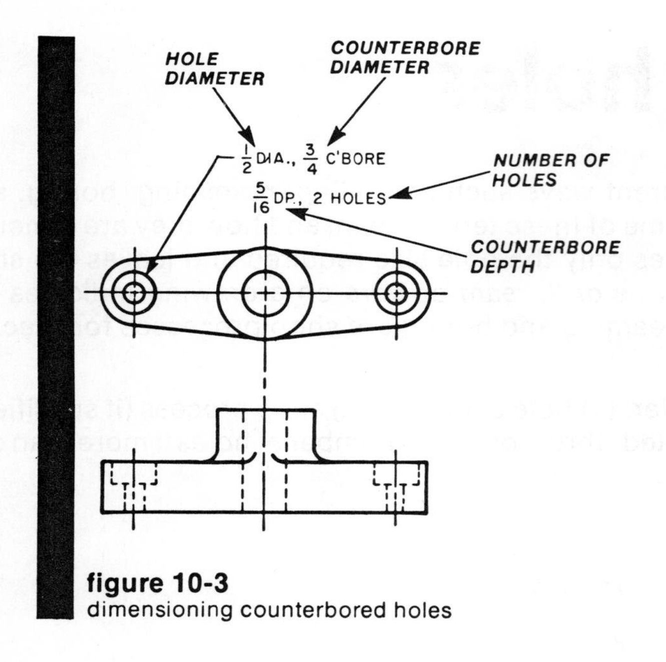

figure 10-3
dimensioning counterbored holes

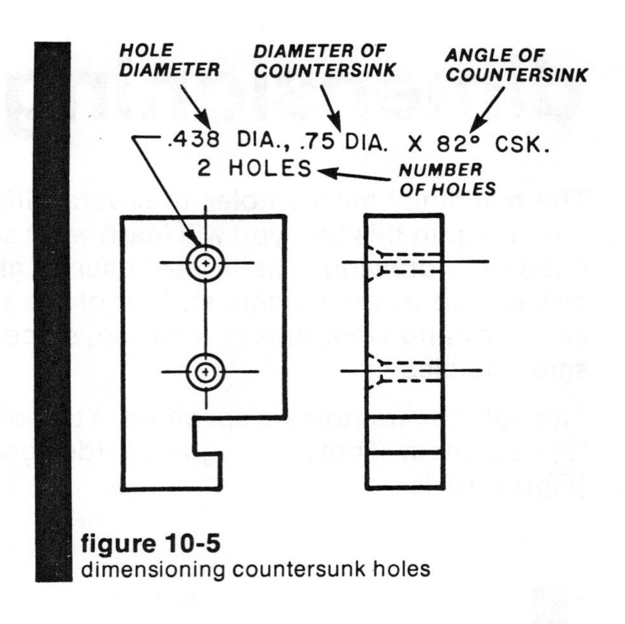

figure 10-5
dimensioning countersunk holes

countersunk holes

A countersink (Figure 10-4) is a conical recess for a flat head machine screw. The notation for a countersunk hole is specified in this order: (1) hole diameter (and depth if not a through-hole), (2) diameter of countersink, (3) angle of countersink, (4) number of holes if more than one (Figure 10-5).

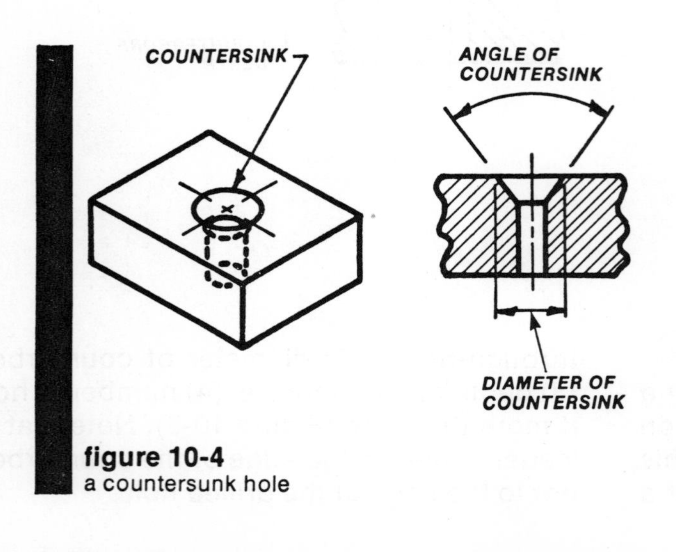

figure 10-4
a countersunk hole

spotfaced holes

A spotface (Figure 10-6) is a smooth, circular surface on a part. A spotface provides a flat seat for a nut or bolt. Spotfacing is usually done on castings and other rough-surfaced parts. The notation for a spotface is specified in this order: (1) hole diameter (and depth if not a through-hole), (2) diameter of spotface or size and type bolt for which spotface is made (Figure 10-7). Note that the leader terminates on the edge of the spotface. The spotface is drawn 1/16" deep, although the actual depth necessary to obtain a smooth, flat surface is usually determined by the machinist.

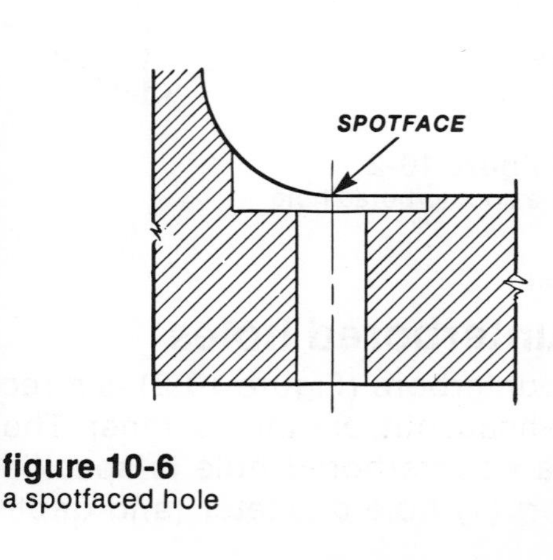

figure 10-6
a spotfaced hole

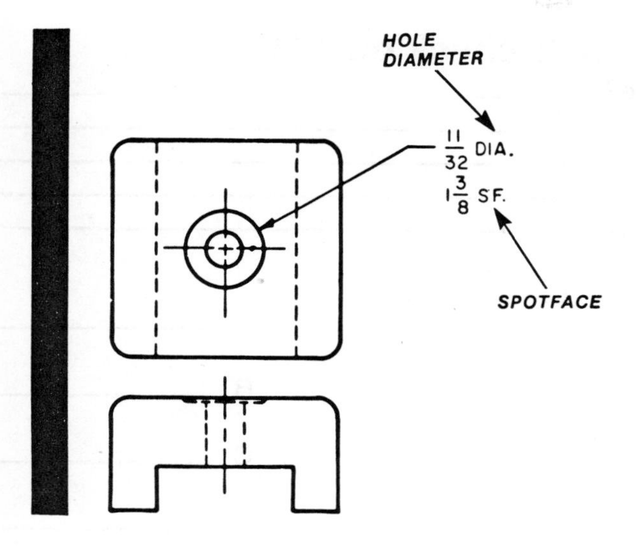

figure 10-7
dimensioning a spotfaced hole

tapped holes

For tapped (threaded) holes, (1) the tap drill size, drill depth (if not a through-hole), (2) the thread specification, and (3) thread depth are given. (Thread specifications and thread representation are covered in detail in Unit 14.) The leader terminates on the hidden line which represents the root of the threads (Figure 10-8). The drafter sometimes specifies the thread size only, and leaves the tap drill selection to the machinist.

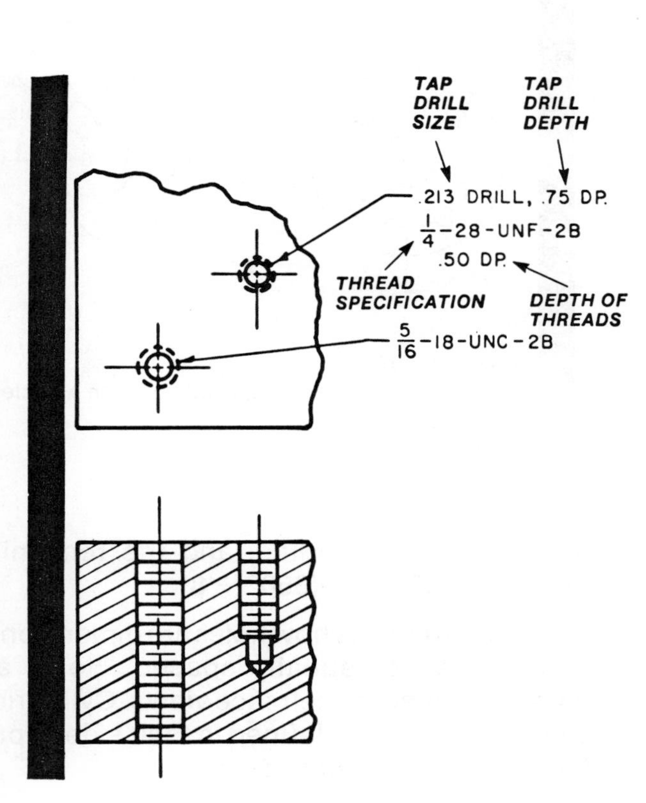

figure 10-8
dimensioning tapped holes

self-check quiz 10-a

1. What is the size of the plain, drilled hole? ____________
2. What is the size of the counterbore? ____________
3. How deep is the counterbore? ____________
4. What is the angle of the countersink? ____________
5. What is the diameter of the spotface? ____________
6. Determine the following dimensions:

$\frac{1}{2}$ DIA.

$\frac{1}{4}$ DIA., $\frac{5}{8}$ CB. $\frac{5}{16}$ DP.

$\frac{1}{4}$-20-UNC-3B

$\frac{5}{16}$ DIA. $\frac{21}{32}$ S'FACE

$\frac{5}{16}$ DIA. $\frac{5}{8}$ DIA. X 90° CSK.

A ____________

B ____________

C ____________

D ____________

E ____________

dimensioning holes on a circle

Holes which are equally spaced on a circle are normally located in relation to a centerline of the object (Figure 10-9). The diameter of the circle on which the holes are located is called the *bolt circle diameter*. In Figure 10-9, the first hole is on a centerline, and the remaining holes are 45° apart (360° ÷ 8).

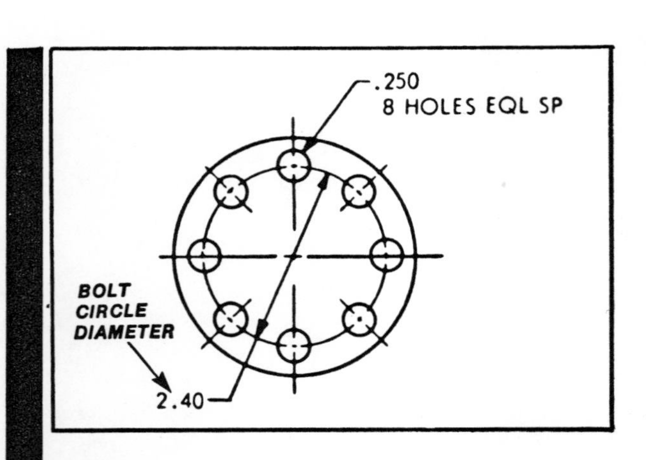

figure 10-9
holes equally spaced on a circle (ANSI Y14.5)

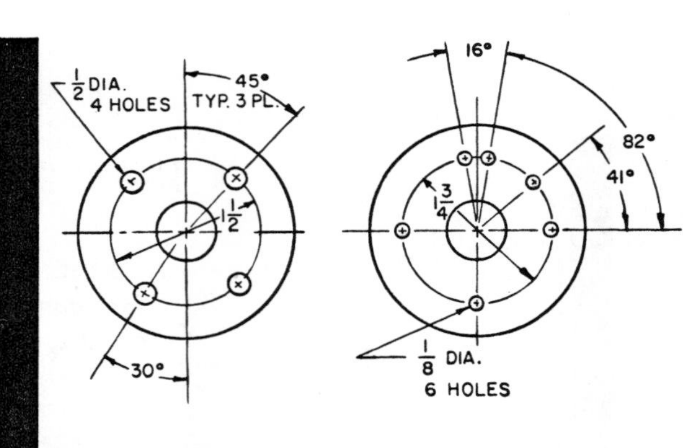

figure 10-10
dimensioning unequally spaced holes on a circle

Figure 10-10 shows how holes are dimensioned if they are not equally spaced. Holes are usually defined in their face views. Cylindrical shapes are usually defined in their side or profile views. See Figure 10-11.

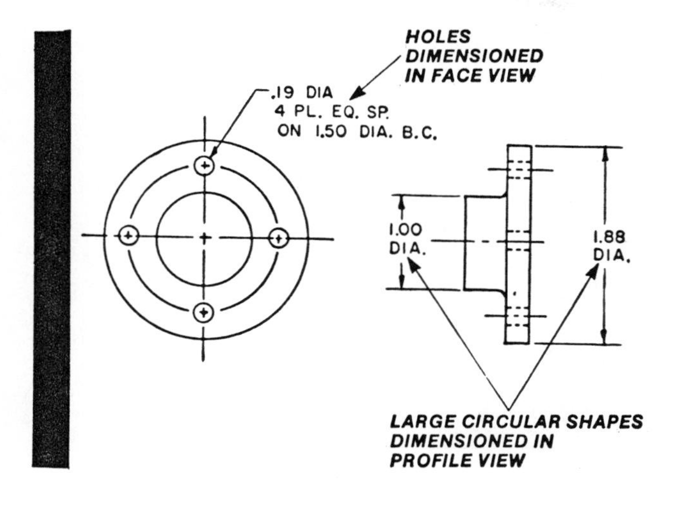

figure 10-11
proper location of dimensions

dimensioning holes and surfaces from a baseline

When a part contains several holes, the drafter sometimes uses a technique known as *baseline,* or *datum*, dimensioning.

Figure 10-12 shows an object with all features dimensioned from the bottom and right side edges. These edges are called *datum planes*. If the object is round, the datum planes are usually taken as centerlines of the object (Figure 10-13).

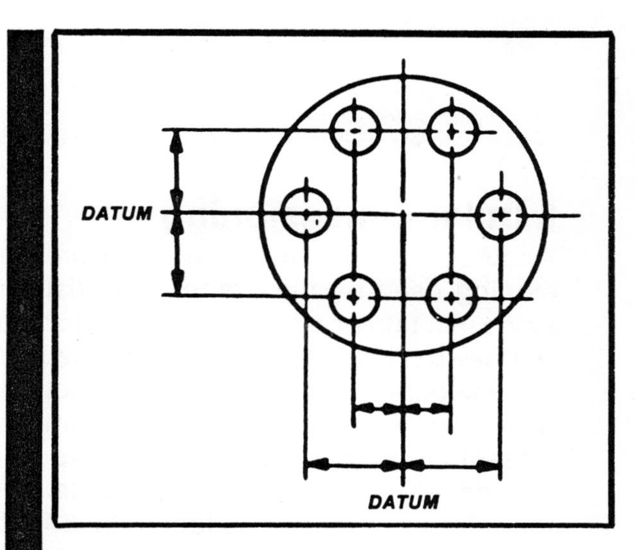

figure 10-13
a round object with centerlines as datum planes (ANSI Y14.5)

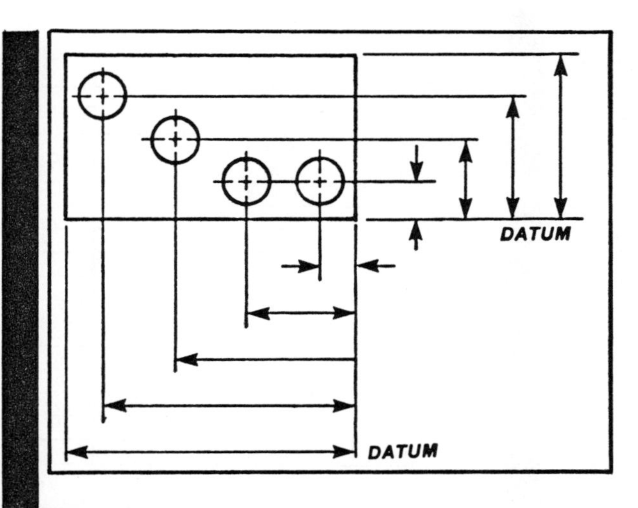

figure 10-12
an example of datum (base line) dimensioning

If the datum planes are indicated as zero coordinates, dimensions from them may be shown on extension lines without the use of dimension lines or arrowheads (Figure 10-14). This is called *ordinate dimensioning.*

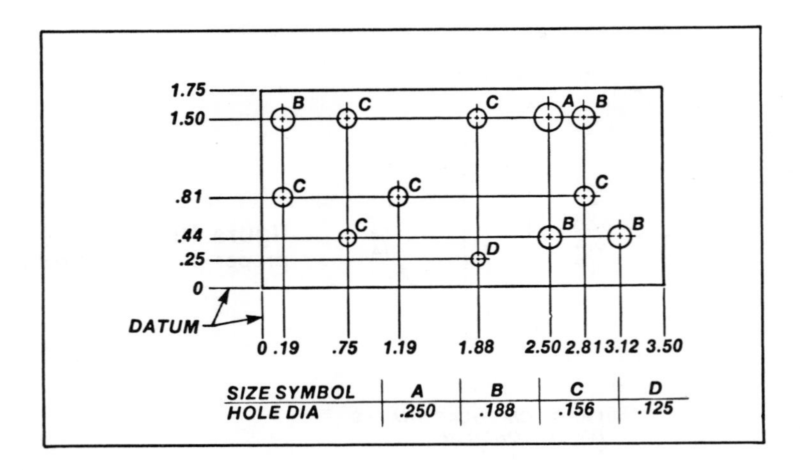

figure 10-14
ordinate dimensioning eliminates dimension lines (ANSI Y14.5)

tabular dimensioning

Another common practice of datum dimensioning used by the drafter is to list dimensions in a table on the drawing rather than on the views (Figure 10-15). This is called *tabular dimensioning,* and is used on objects which have a large number of holes or similar features. In Figure 10-15, to locate hole C6, find C6 under "Hole Symbol" on the table. Read to the left to see that C6 is located .75 from the left edge (X) and .44 up from the bottom edge (Y). It is a .156 diameter hole. Hole B4 is a .188 diameter hole located 3.00 from the left edge (X) and .44 up from the bottom edge (Y).

Many machining operations are now performed by production machines that follow directions given on punched tapes or are electronically controlled. Tabular dimensioning is needed for such numerically controlled machine tools.

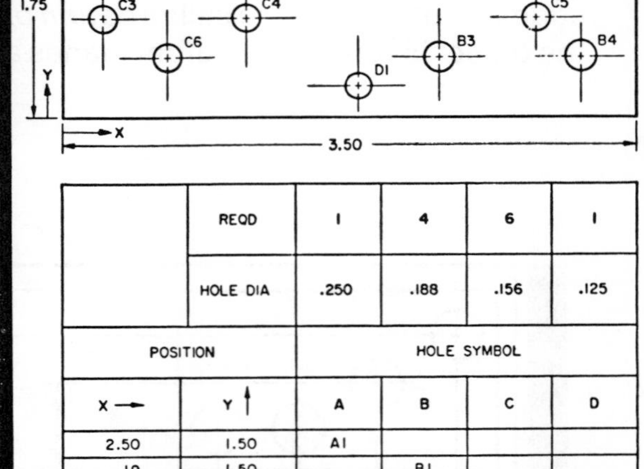

	REQD	1	4	6	1
	HOLE DIA	.250	.188	.156	.125
POSITION		HOLE SYMBOL			
X →	Y ↑	A	B	C	D
2.50	1.50	A1			
.19	1.50		B1		
2.81	1.50		B2		
2.50	.44		B3		
3.00	.44		B4		
.75	1.50			C1	
1.88	1.50			C2	
.19	.81			C3	
1.19	.81			C4	
2.81	.81			C5	
.75	.44			C6	
1.88	.25				D1

figure 10-15
tabular dimensioning (ANSI Y14.5)

self-check quiz 10-b

1. Which lines on the figure establish the datum planes?

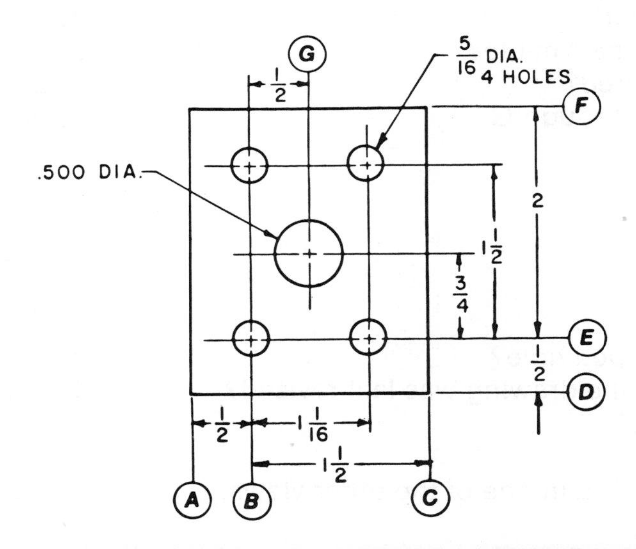

1. ________________

exercise 10-1

______________ name

1. What is the bolt circle diameter of the 7/16 DRILL Holes?
2. What is the diameter of the spotface of the 7/16 DRILL holes?
3. What is the depth of the spotface of the 7/16 DRILL holes?
4. How deep is the counterbore of the 1/2 DIA holes?
5. Name all the different machining operations noted.

1. ________________
2. ________________
3. ________________
4. ________________
5. ________________

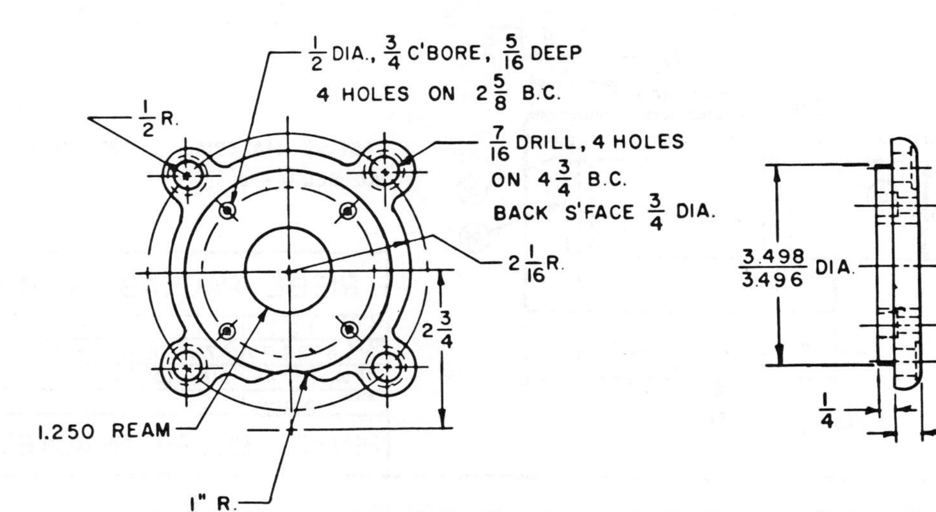

exercise 10-2

______________________ name

1. What is the name of the part? 1. ________
2. From what material is the part made? 2. ________
3. How many parts are required? 3. ________
4. Determine dimensions A through G. 4. A ________ B ________ C ________ D ________ E ________ F ________ G ________
5. What is the size of the tapped hole? 5. ________
6. What was changed when the drawing was last revised? 6. ________
7. What scale is the drawing? 7. ________
8. Name the views shown. 8. ________
9. Identify surfaces H through L in one of the other views. 9. ________

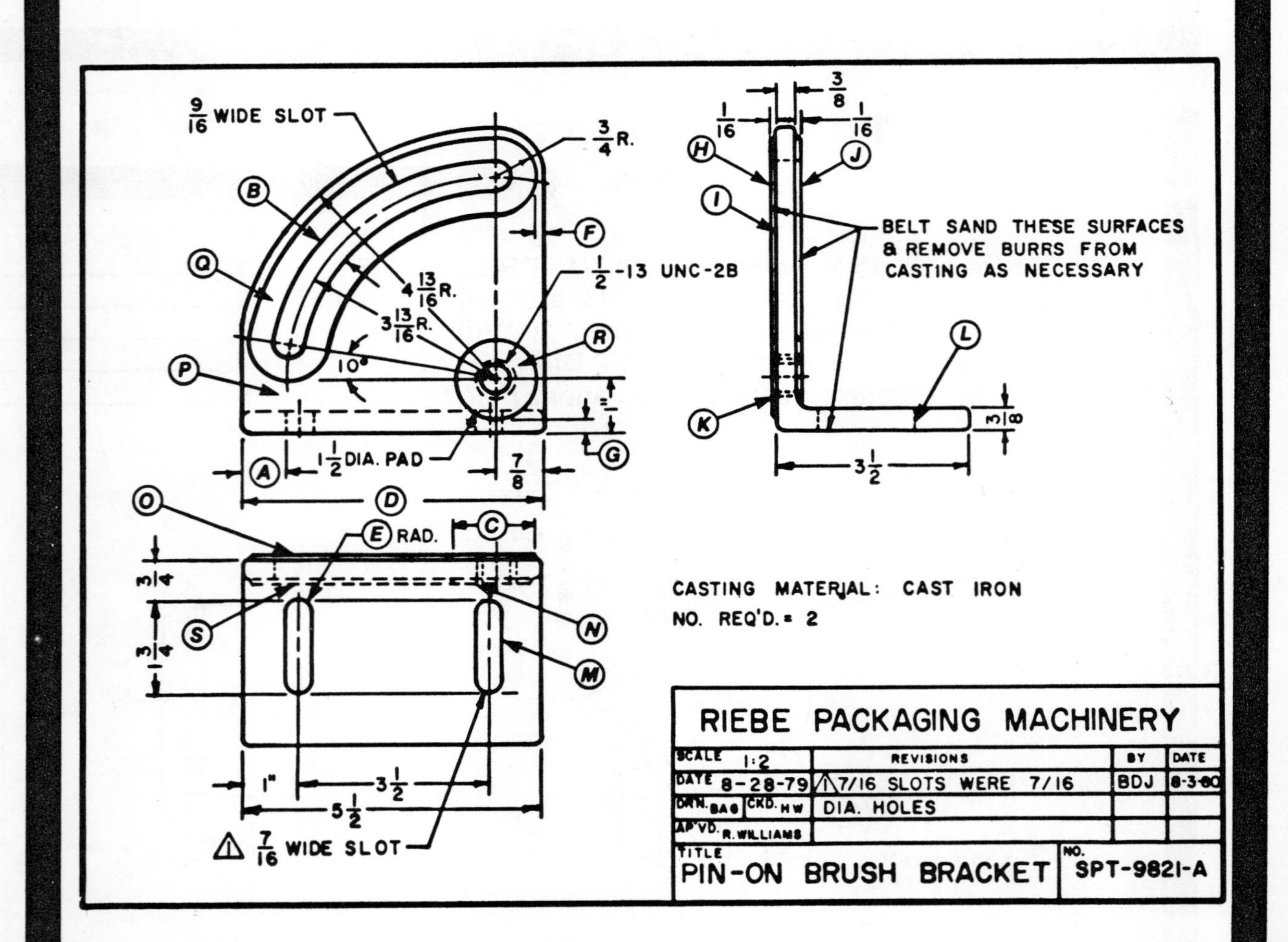

exercise 10-3

________________________ name

1. Determine dimensions A through N
2. How many plates are required?
3. What material is to be used for the plate?
4. What material is to be used for the keystock?
5. How long is the piece of keystock?

1. A ____
   B ____
   C ____
   D ____
   E ____
   F ____
   G ____
   H ____
   I ____
   J ____
   K ____
   L ____
   M ____
   N ____
2. ____
3. ____
4. ____
5. ____

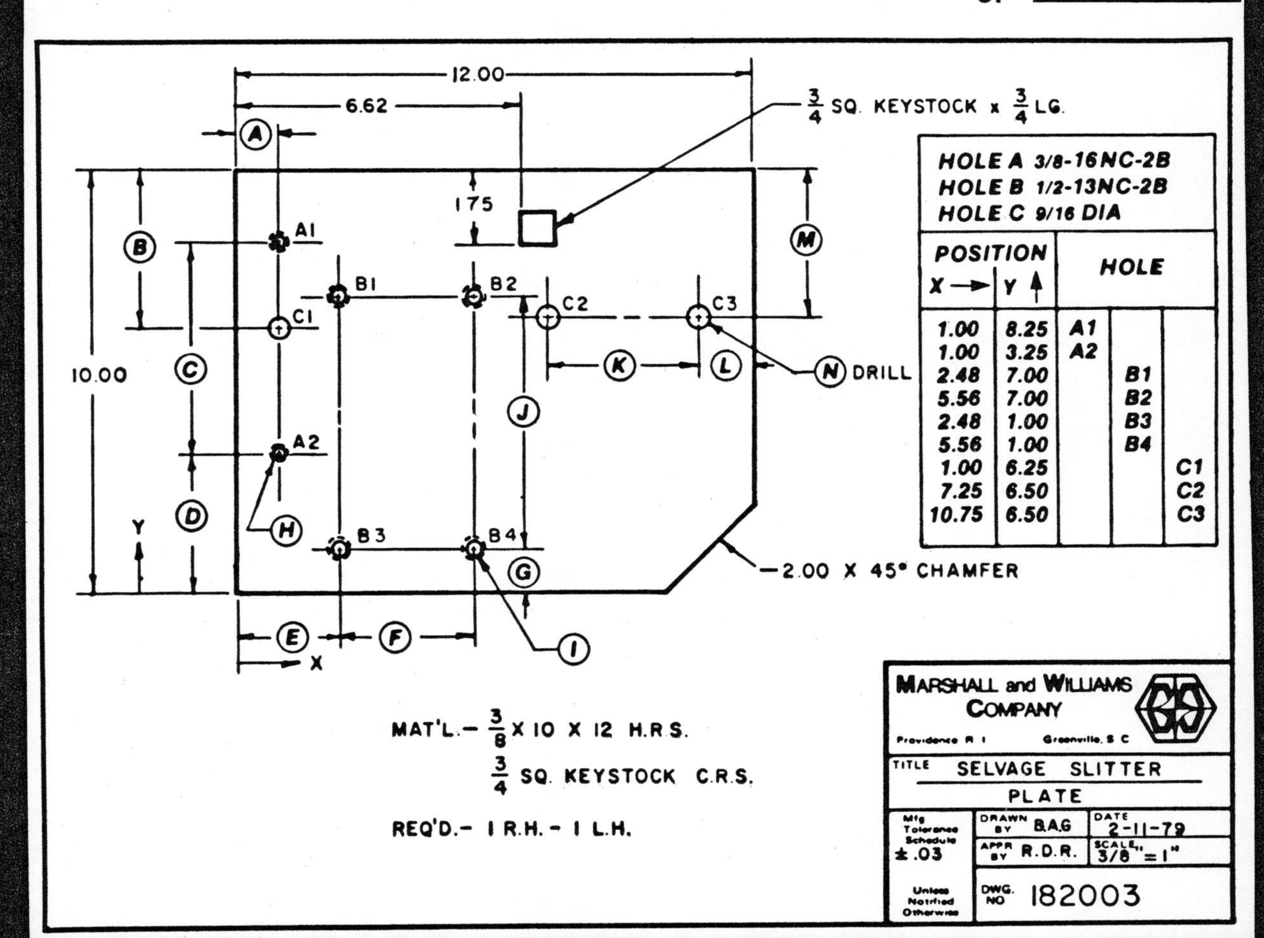

This page for student notes

limits and tolerances

Tolerances are permissible variations in size and location dimensions. You cannot machine parts to exact size every time. The designer determines the degree of machining accuracy necessary by considering the following factors:

1. The fit of mating parts (how the parts will go together).
2. The material of the parts.
3. The permissible cost of manufacturing.
4. The quantity to be made.

These factors establish the tolerances (limits of accuracy) for dimensions. After completing this unit you will be able to recognize the different kinds of tolerances. You will be able to apply to tolerances to machining dimensions.

decimal tolerances

Decimal tolerances are expressed in thousandths of an inch. If the tolerance of a dimension is given as ± .030 (plus or minus 30 thousandths), this means that the dimension is allowed to be .030 more than or .030 less than the nominal value. Thus in the following example, the largest permissible size is

2.00 + .030 = 2.03

± .030
2.00

and the smallest permissible size is

2.00 - .030 = 1.97.

The largest permissible size is called the *upper limit*. The smallest permissible size is called the *lower limit*. Any value between these two limits is acceptable.

fractional and angular tolerances

Fractional and angular tolerances are expressed in a similar way. Figure 11-1 shows a part with decimal, fractional and angular tolerances.

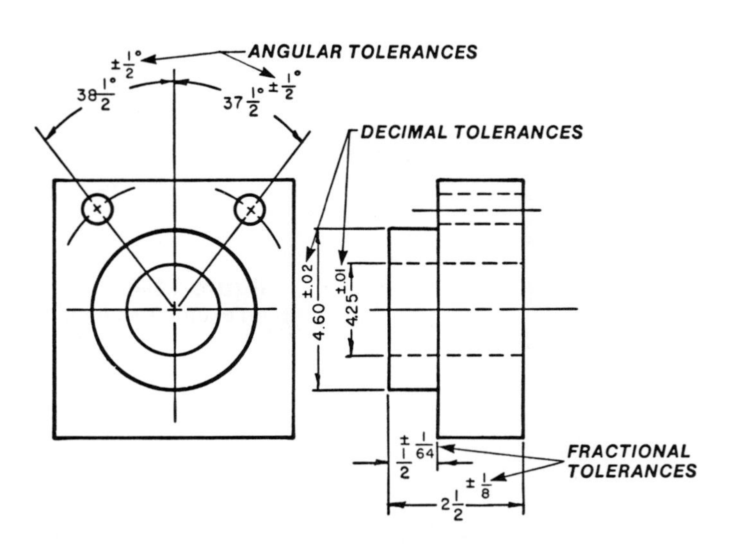

figure 11-1
a drawing with decimal, fractional and angular tolerances

unilateral and bilateral tolerances

The designer expresses tolerances in accordance with the intended fit and function of the part. Tolerances considered thus far have been *bilateral tolerances*. This means that tolerances apply above and below the nominal dimension. The bilateral tolerances have also been of equal value above and below the nominal dimension. Some bilateral tolerances, however, are not equal (Figure 11-2).

Sometimes tolerances are *unilateral*. That means they apply in one direction (plus or minus) only (Figure 11-3).

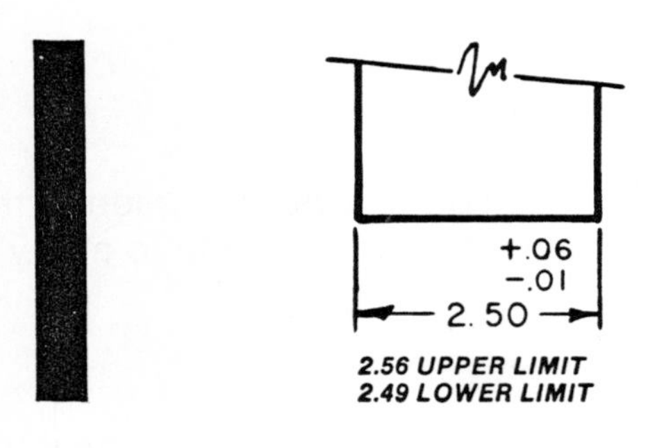

figure 11-2
an example of a bilateral tolerance which is unequal

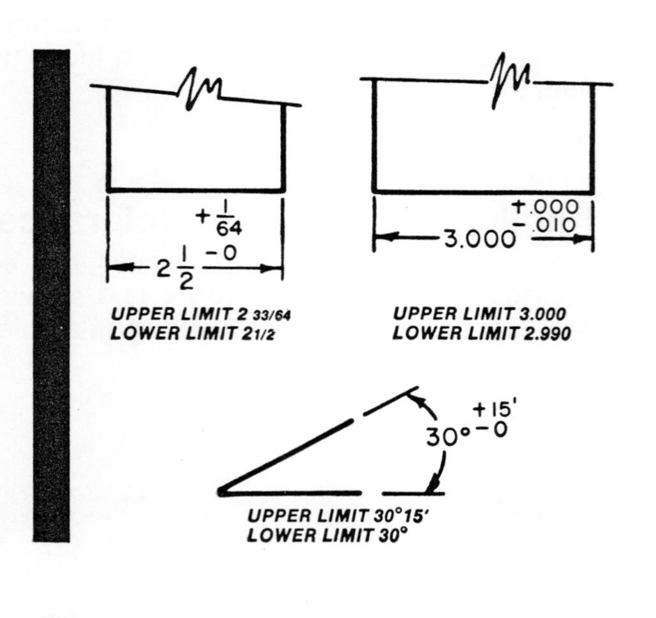

figure 11-3
examples of unilateral (one-sided) tolerances

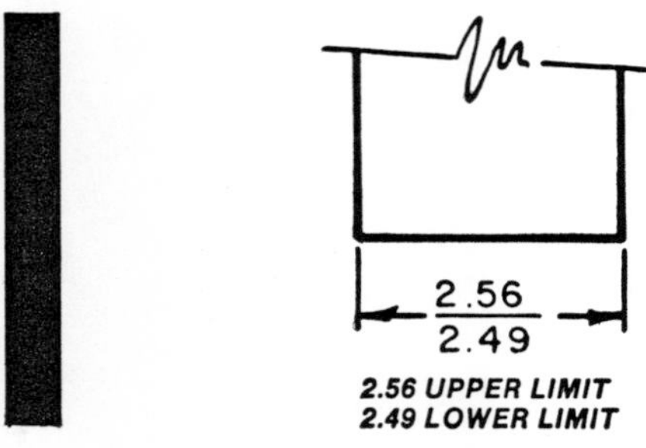

figure 11-4
an example of limit dimensioning

A dimension expressed by its upper and lower limits directly on a drawing is known as *limit dimensioning* (Figure 11-4).

self-check quiz 11a

List the dimensions with corresponding tolerances

1. Bilateral equal tolerances ____________________
2. Bilateral unequal tolerances ____________________
3. Unilateral tolerances ____________________
4. Limit dimensions ____________________
5. Size dimensions ____________________

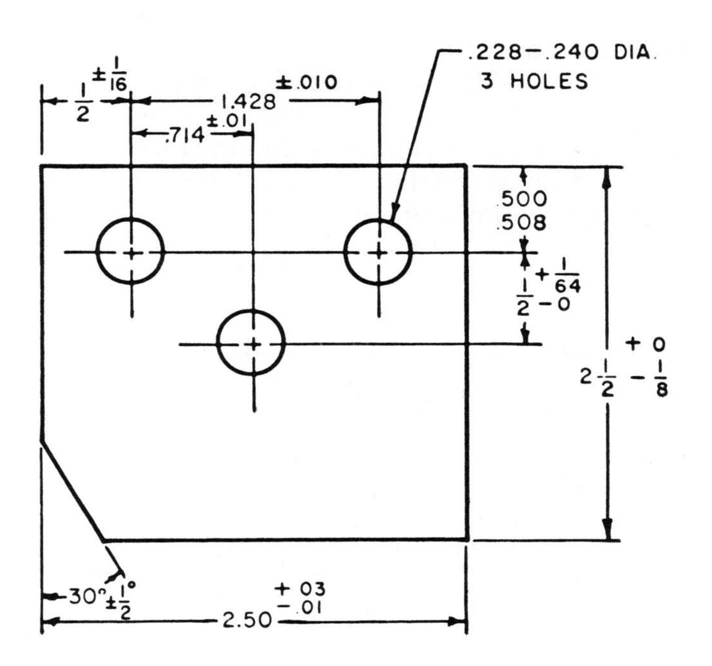

exercise 11-1

______________________________ name

1. What is the scale of the drawing? 1. __________
2. What material is the bushing made from? 2. __________
3. What was the original outside diameter of the bushing? 3. __________
4. What is the latest revision of the drawing? 4. __________
5. What is the upper limit of the bore? 5. __________
6. What is the lower limit of the outside diameter? 6. __________
7. What type of dimensioning is used to express the bore dimension? 7. __________
8. Is the tolerance on the 45° dimension unilateral or bilateral? 8. __________
9. Is the tolerance on the 15° dimension unilateral or bilateral? 9. __________
10. Determine the nominal length of the bushing. 10. __________
11. Express the bushing length with bilateral tolerances. 11. __________
12. What is the maximum size of the 1/16 dimension? 12. __________
13. What is the minimum wall thickness, T, of the bushing? 13. __________
14. Determine the maximum value for dimension L. 14. __________

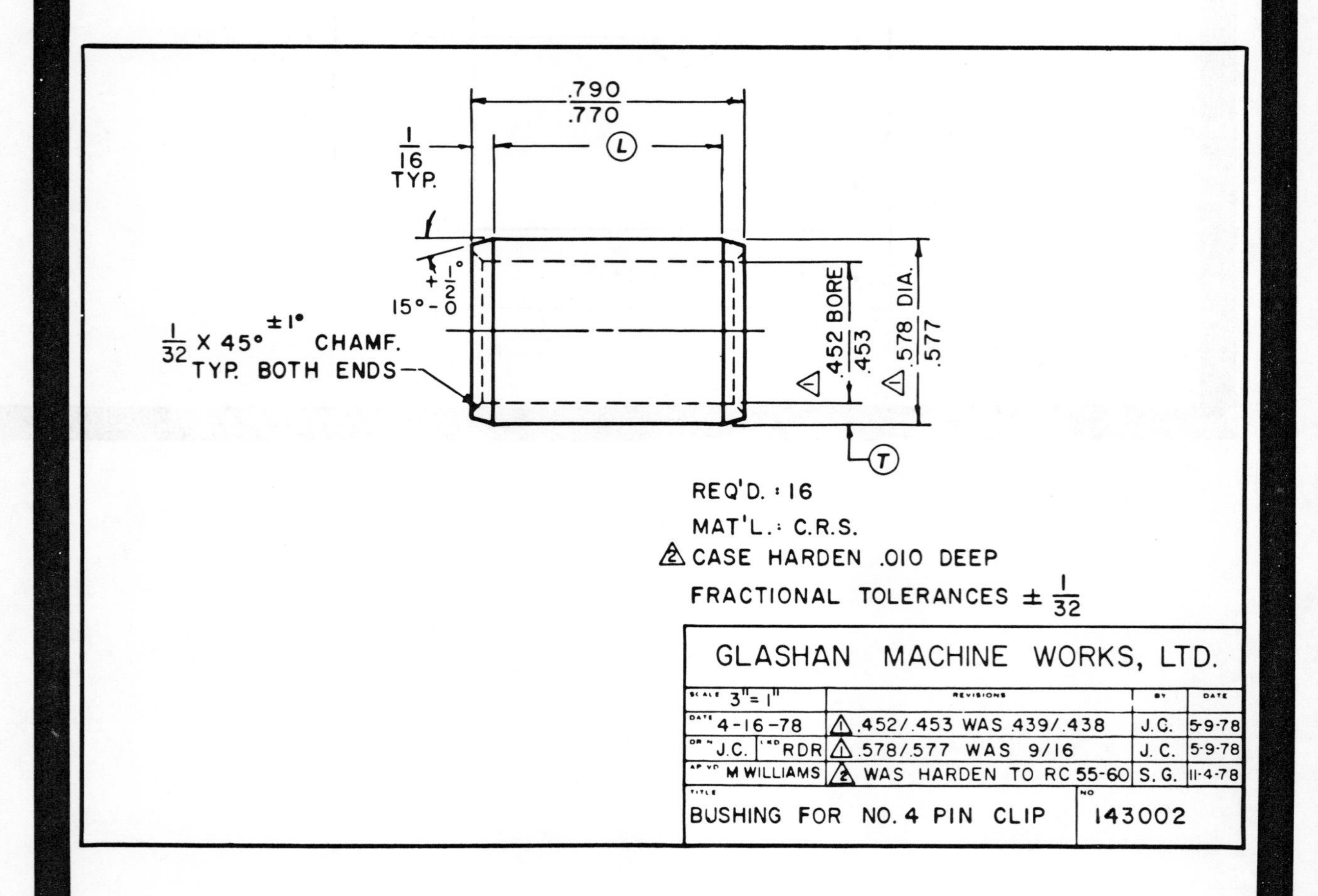

section views

Unit 12

The drafter sometimes draws an internal view to explain hidden features of an object. This helps to understand a complicated object better. We call an internal view a *section view* or a *cross section*. After completing this unit you will be able to recognize and interpret the different kinds of section views. A section view is obtained by passing an imaginary cutting plane through the part. The portion of the object in front of the cutting plane is mentally removed, and what remains is exposed to view in the cross section (Figure 12-1).

Compare the two side views of Figure 12-2. Observe how the section view is much clearer than the regular side view. The section shown is called a *full section* since it passes completely through the object. A section allows you to see the internal construction of an object.

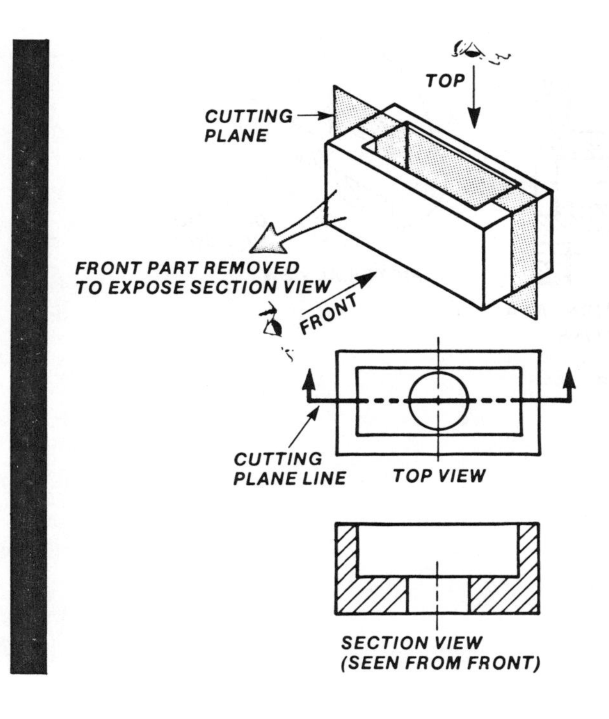

figure 12-1
an imaginary cut through the object exposes the cross-section. Arrows of the cutting plane give the direction of sight for the section view.

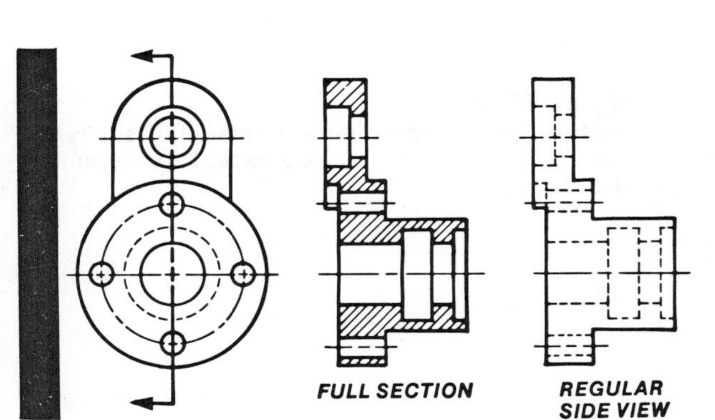

figure 12-2
a section view makes print reading easier. You can see the internal construction of the part.

cutting planes and section lines

The main view from which the section is taken contains a cutting plane line (Figure 12-3). Arrowheads on the cutting plane line indicate the direction of viewing. Letters or numbers identify the section. Identifying letters or numbers are often omitted if it is obvious what the section view represents.

Section lines (crosshatch lines) identify the exposed surfaces which have been "cut" by the cutting plane line. Section lines on a surface all slant in the same direction.

Figure 12-4 shows different section lines which identify the material of the part being sectioned. Because there are so many various materials available today, the drafter usually applies the general purpose (cast-iron) crosshatch lines for a drawing of a single part. Hidden lines are usually omitted in section views, unless they are needed for clarity.

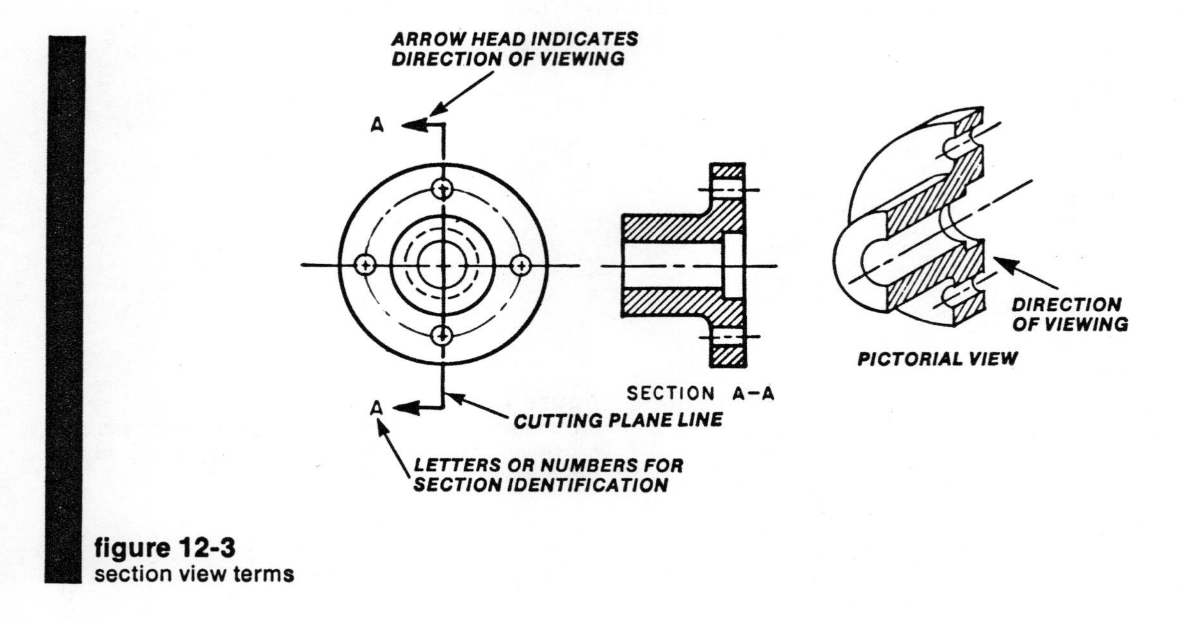

figure 12-3
section view terms

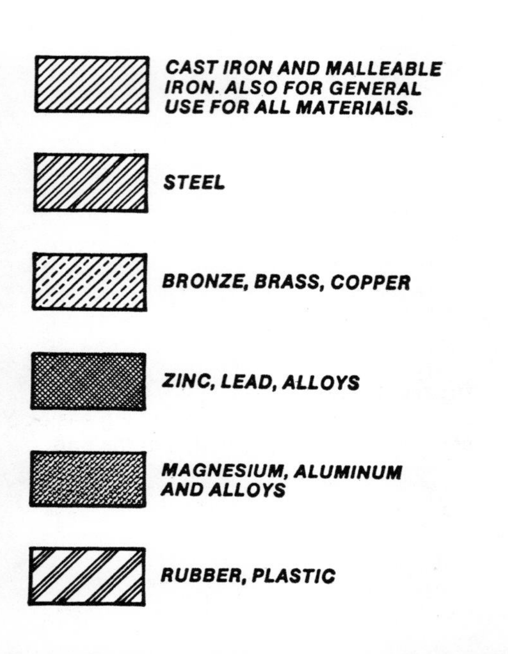

figure 12-4
section lines indicate different materials. Usually the symbol for cast iron is used except for assembly drawings.

self-check quiz 12-a

1. Select the correct section view:

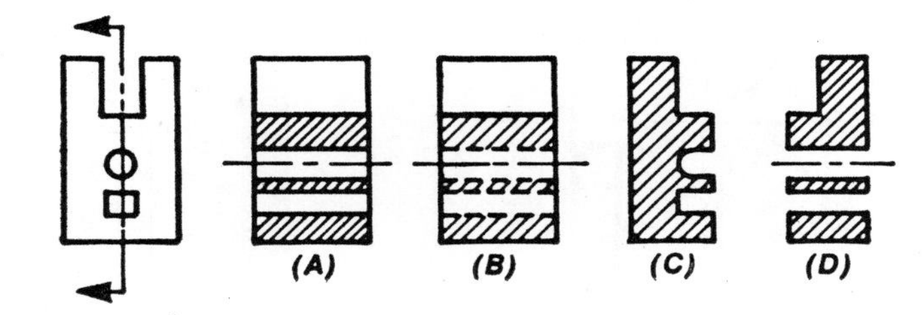

2. Sketch an isometric section view of the object above like the pictorial of figure 12-3.

half sections and partial sections

The drafter may clarify an object by drawing a *half section* combined with half an exterior view (Figure 12-5). Sometimes only a *partial section* or *broken out* section is shown (Figure 12-6).

In both a half section and a partial section, the section view serves a double purpose of showing exterior features plus clarifying some internal details. If it is obvious how a section view is taken, the drafter may omit the cutting plane line from the main view (Figure 12-7).

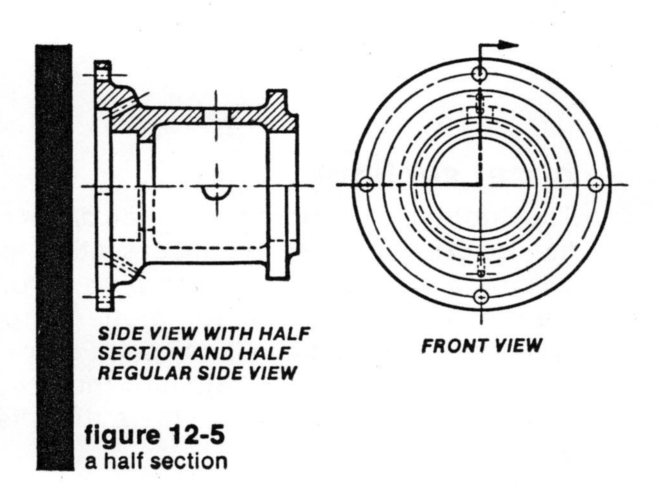

figure 12-5
a half section

figure 12-6
a partial (broken out) section

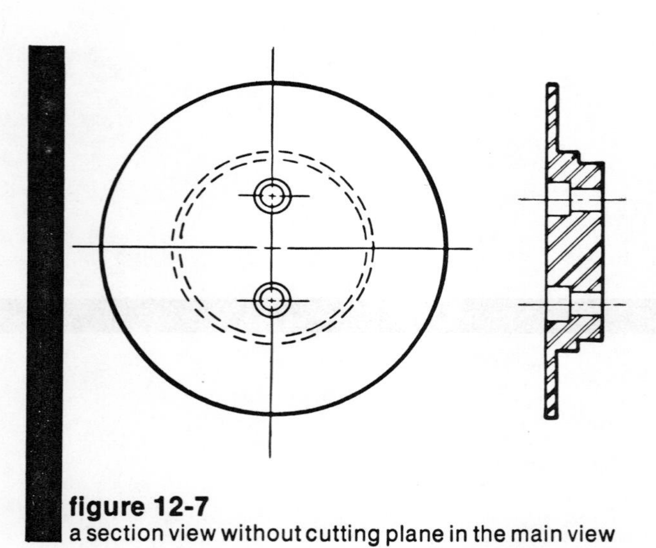

figure 12-7
a section view without cutting plane in the main view

offset sections

The drafter may choose to show an *offset section* for greater clarity. An offset section shifts or moves within the object to pick up other features (Figure 12-8A). The drafter actually shows the cutting plane line as in Figure 12-8B. Notice in Figure 12-8C that the section view does not have breaks where the offsets occur.

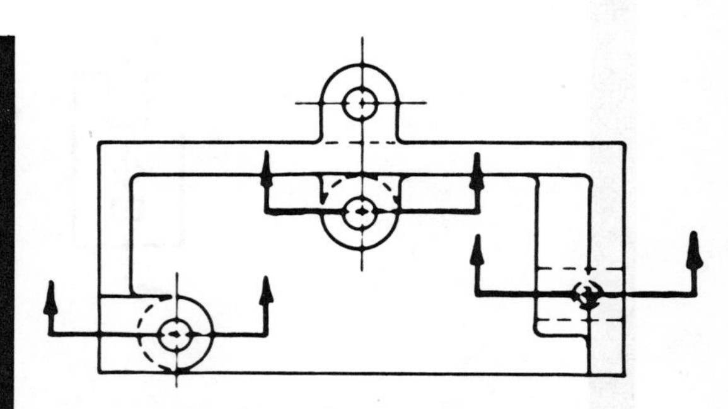

(A) TOP VIEW SHOWING SECTION DESIRED

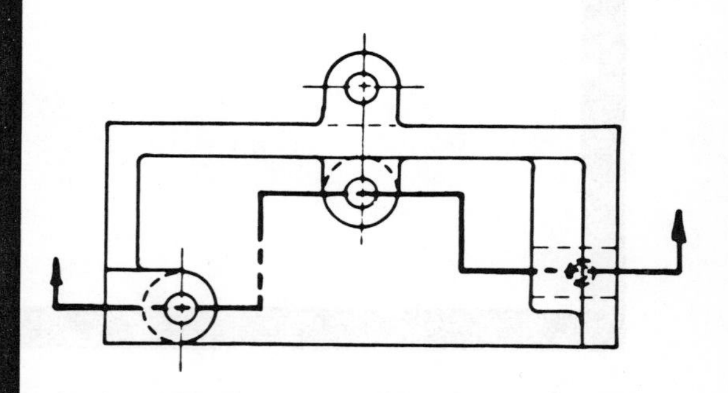

(B) TOP VIEW WITH PROPER CUTTING PLANE LINE

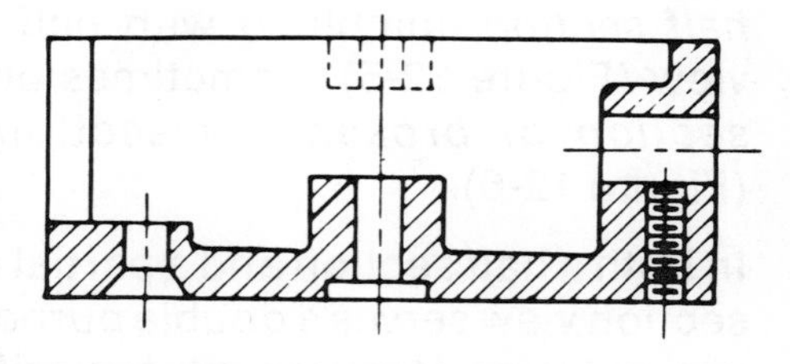

(C) CORRESPONDING SECTION VIEW

figure 12-8
an offset section

self-check quiz 12-b

Determine dimensions A through H.

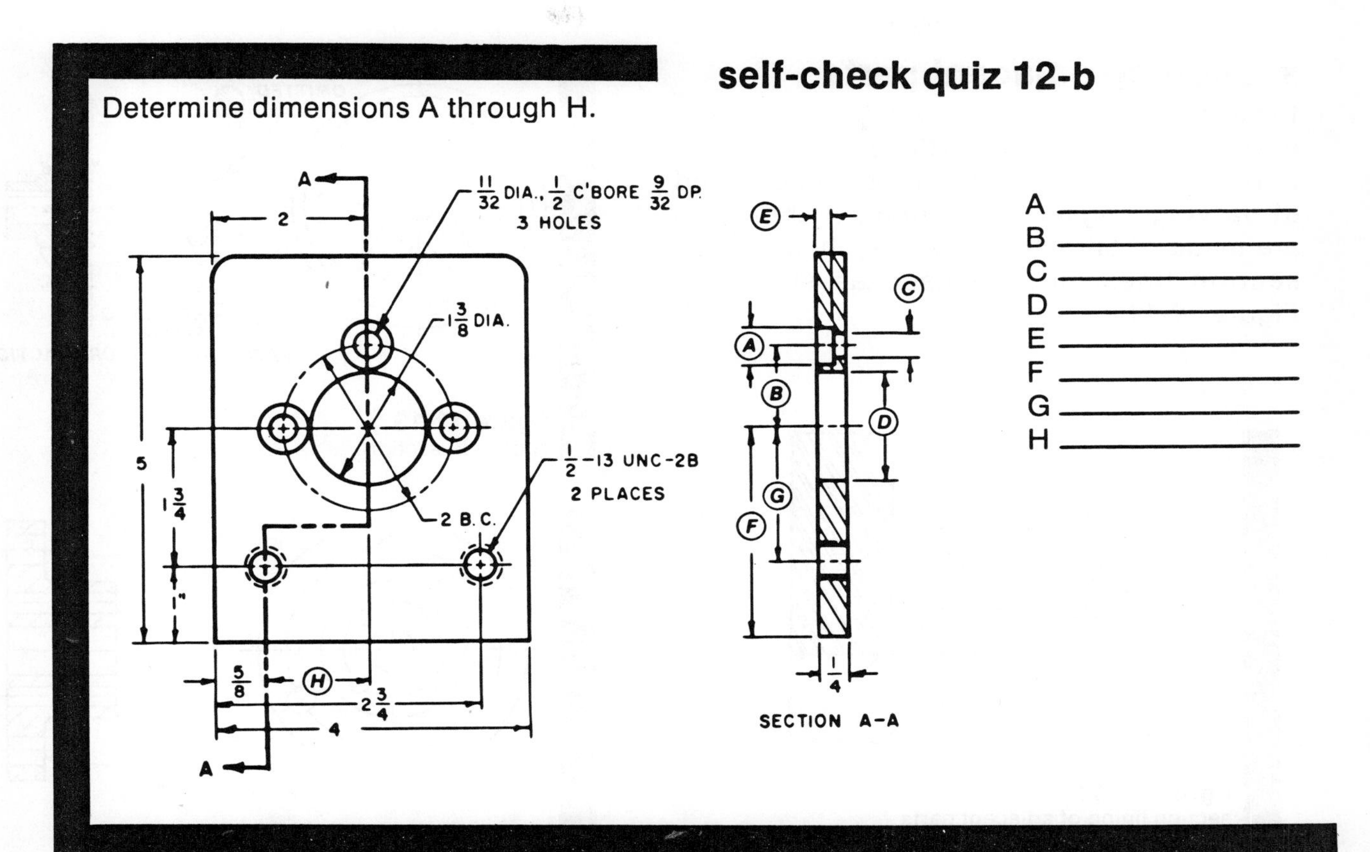

A ______________
B ______________
C ______________
D ______________
E ______________
F ______________
G ______________
H ______________

revolved and removed sections

Sometimes the drafter shows the cross-sectional shape of an object by revolving (turning) the cross-section into the plane of a main view. This is known as a *revolved section*. Figure 12-9 shows three revolved sections on a part. If the revolved cross-section is shown off the part it is called a *removed section* (Figure 12-10).

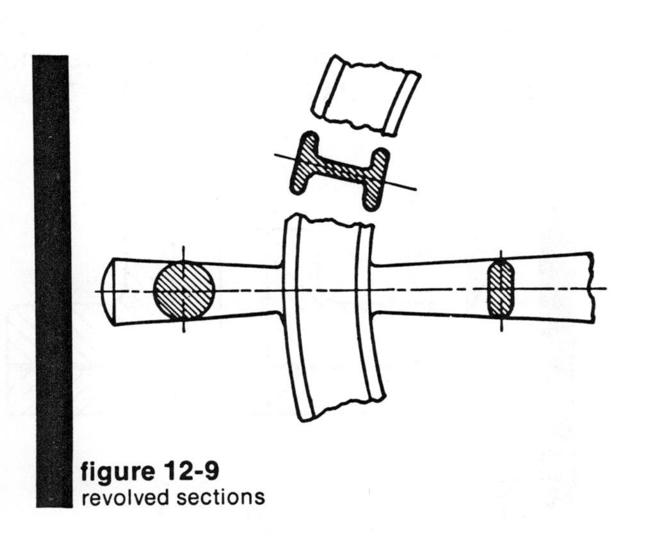

figure 12-9
revolved sections

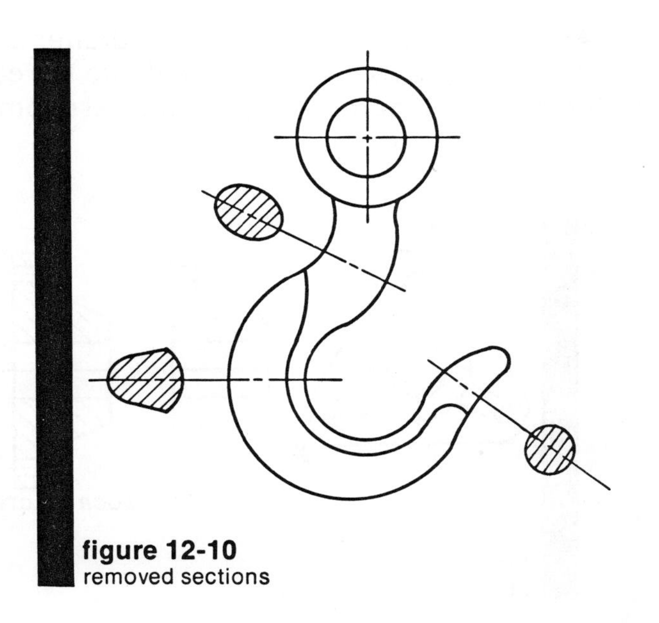

figure 12-10
removed sections

sectioning adjacent parts

If a section view is of an assembly containing more than one part, the mating parts have section lines sloping in different directions and at varying angles. Of course, if the parts are made of different materials, then various section lines may be used as shown in Figure 12-11.

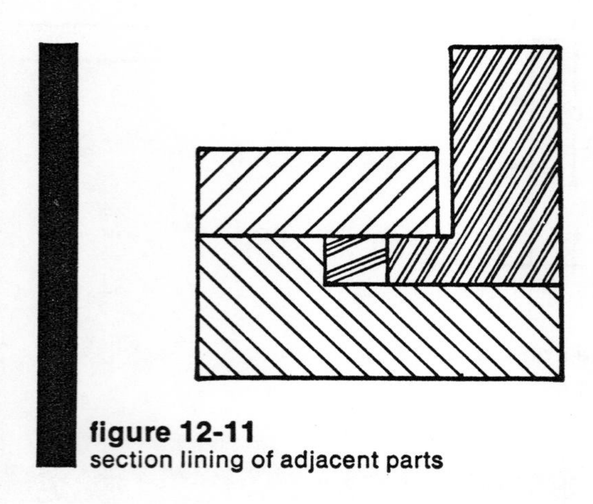

figure 12-11
section lining of adjacent parts

spokes and ribs in section

The drafter usually does not put section lines on spokes or ribs. Figure 12-12 shows why this is so. The sectioned rib gives a false impression of the object.

Another common practice of the drafter is to draw a spoke as if it was rotated into the cutting plane (Figure 12-13). In this case, other

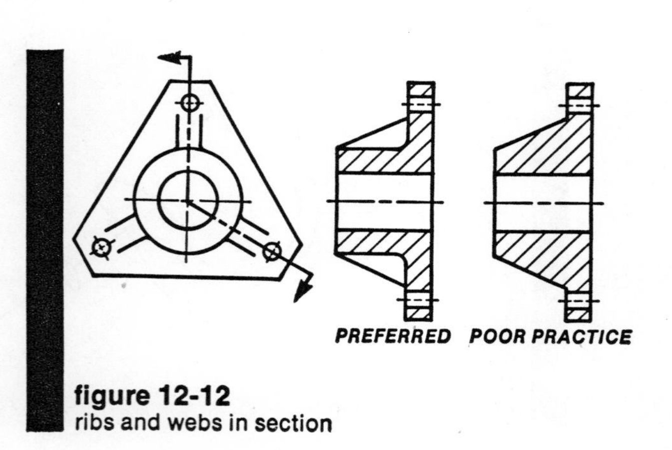

figure 12-12
ribs and webs in section

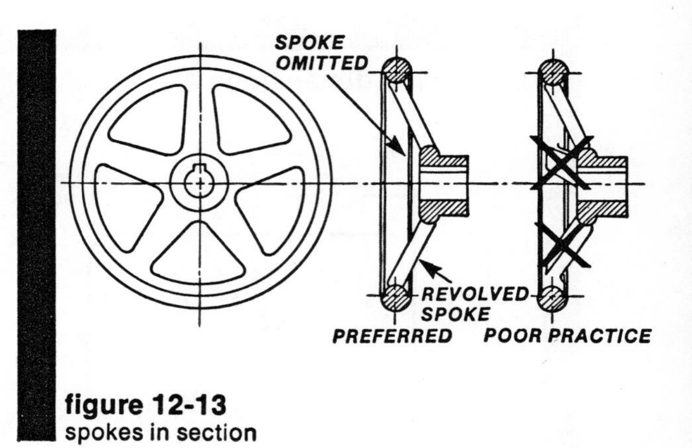

figure 12-13
spokes in section

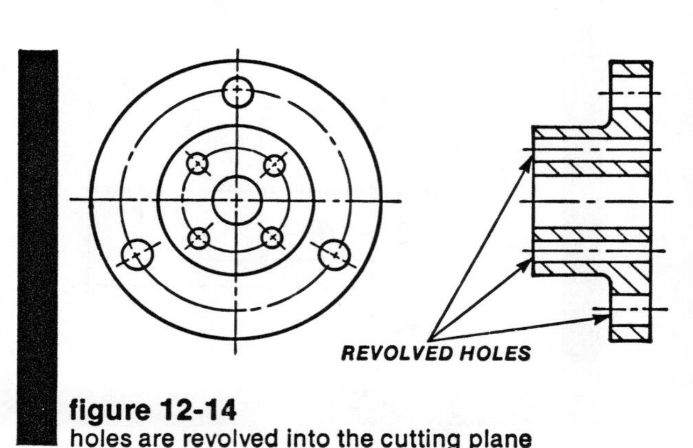

figure 12-14
holes are revolved into the cutting plane

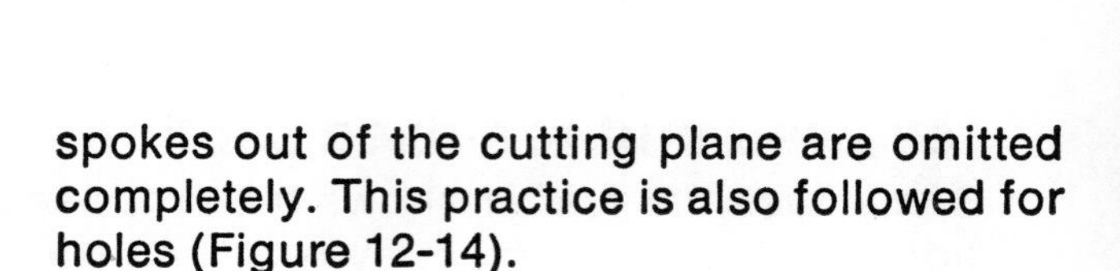

spokes out of the cutting plane are omitted completely. This practice is also followed for holes (Figure 12-14).

thin sections

The drafter shows the cross section solid black for very thin objects such as shims, plates, and gaskets (Figure 12-15).

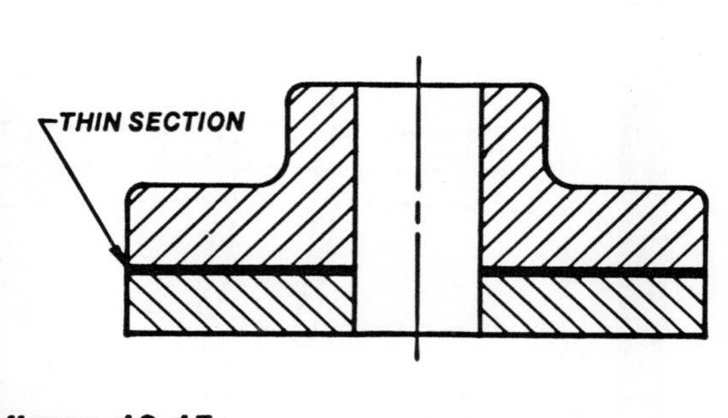

figure 12-15
thin sections are shown solid black

exercise 12-1

________________________ name

Sketch the section view for each object.

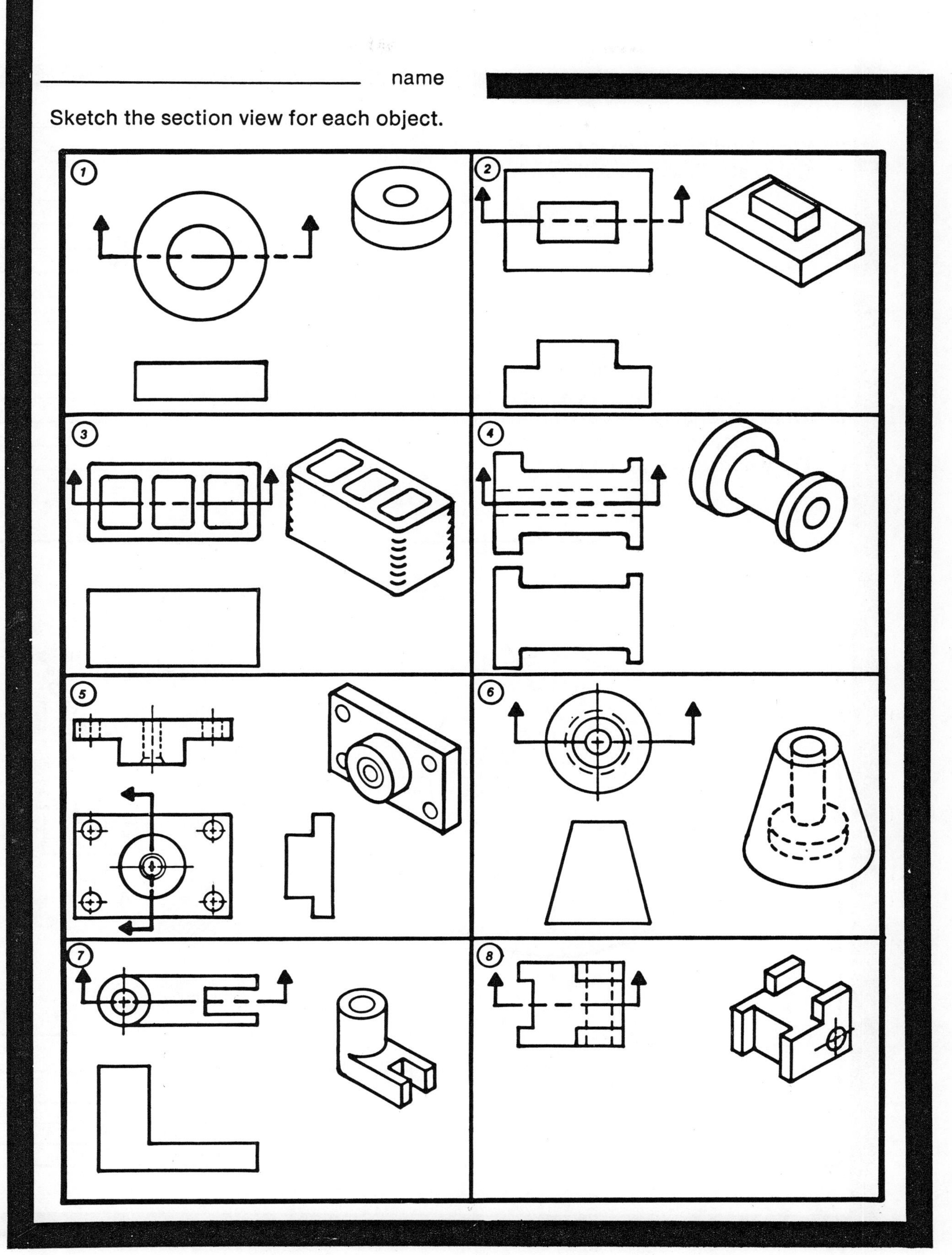

exercise 12-2

____________________ name

1. What is the name of the part? 1. ________
2. What kind of section is shown? 2. ________
3. Calculate distances A through G. 3.
   - A ________
   - B ________
   - C ________
   - D ________
   - E ________
   - F ________
   - G ________
4. What size is the tapped hole? 4. ________
5. How far is the tapped hole from the front face of the handwheel? 5. ________
6. What material is the part made from? 6. ________
7. What casting pattern is used to make the part? 7. ________
8. For whom is the part made? 8. ________

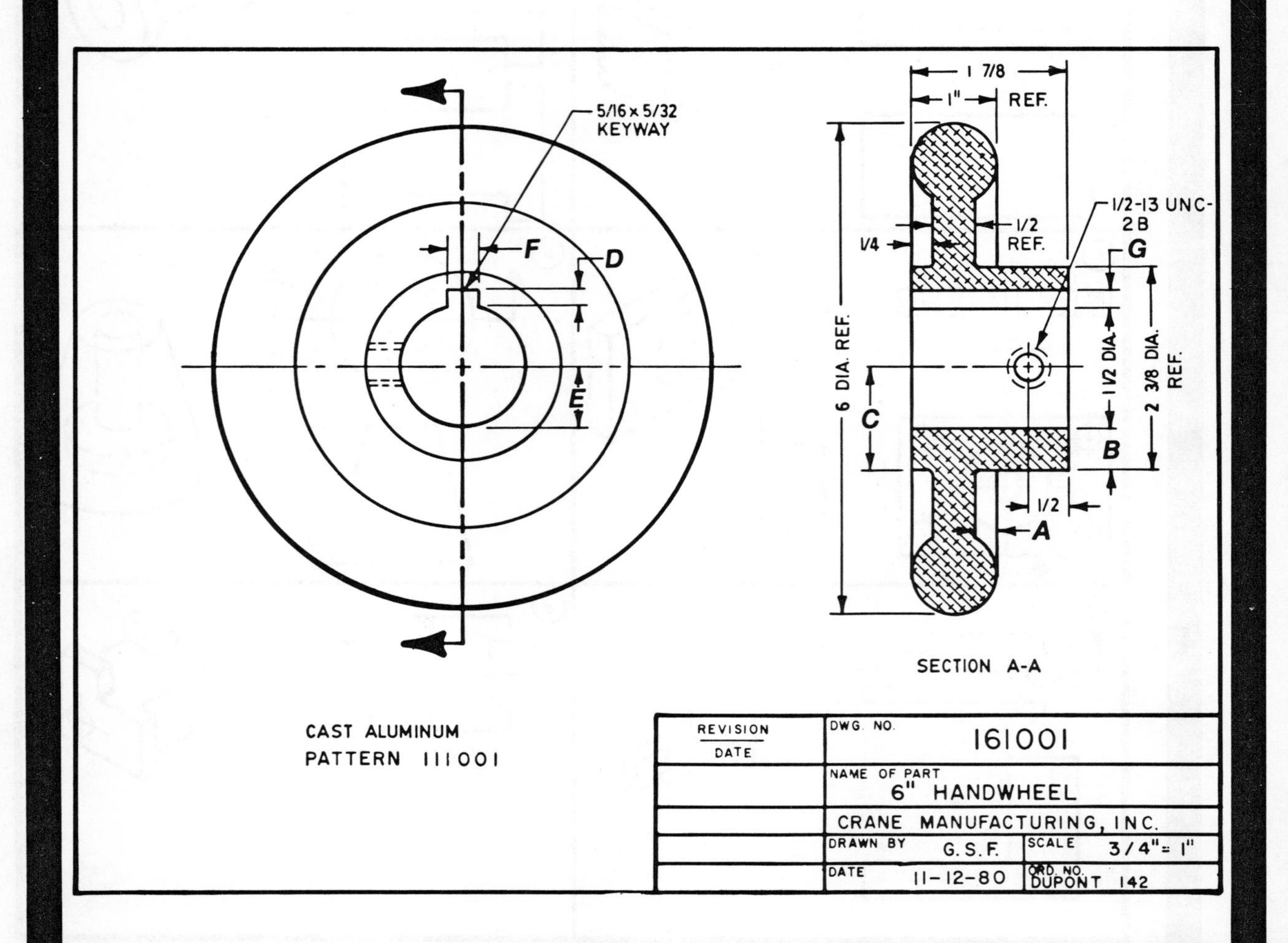

exercise 12-3

______________________ name

1. Determine dimensions A through G.
2. What is the maximum possible depth of the 5.490/5.500 hole?
3. What is the maximum possible depth of the 2.127/2.129 hole?
4. List all reference dimensions.
5. What material is the turret cam follower made of?
6. Determine the minimum possible value for dimension H.
7. Determine the maximum possible value for dimension I.

1.
A ________
B ________
C ________
D ________
E ________
F ________
G ________
2. ________
3. ________
4. ________
5. ________
6. ________
7. ________

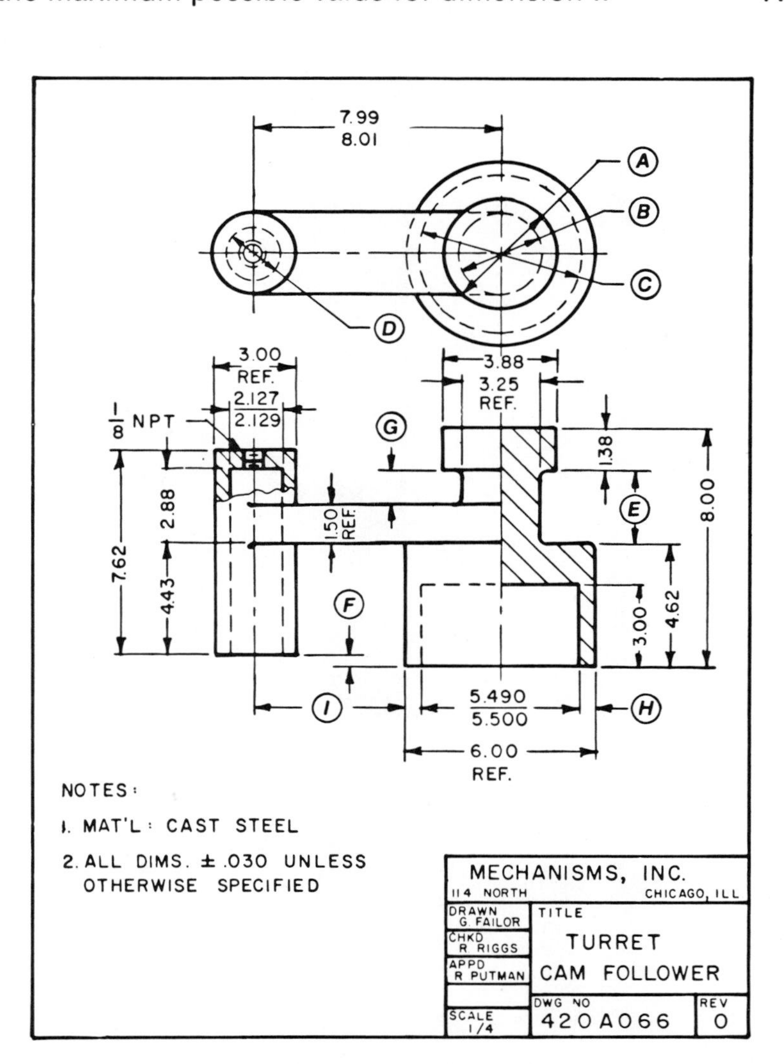

This page for student notes

auxiliary views

So far we have considered objects with surfaces which are either parallel or perpendicular (at a right angle) to one another. However, many objects have surfaces or features that are inclined. An inclined surface appears distorted in the normal views (Figure 13-1). In this unit you will learn how the drafter shows inclined surfaces of objects by auxiliary views.

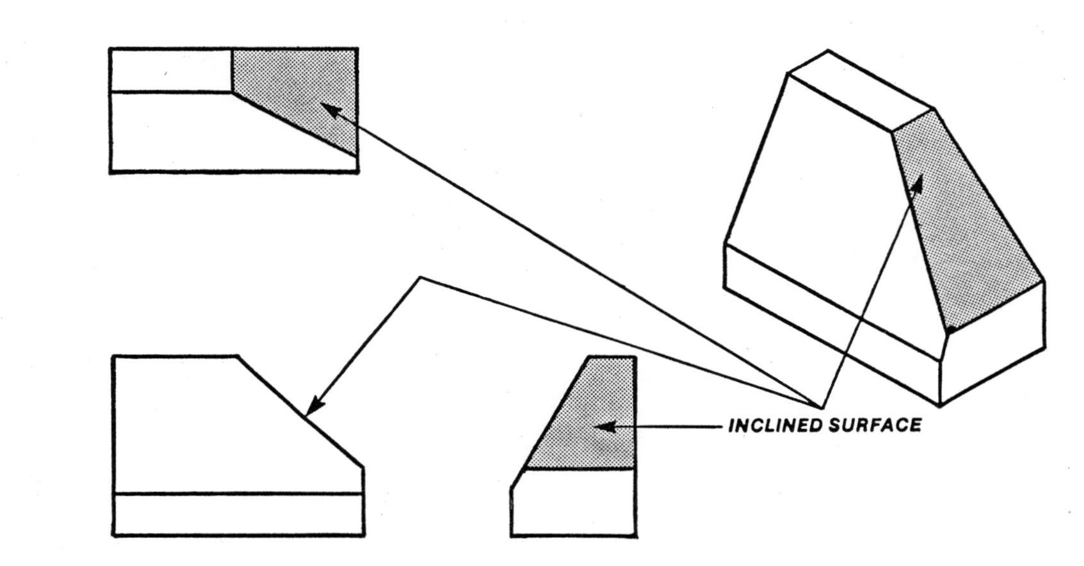

figure 13-1
none of the main views shows the true size and shape of an inclined surface

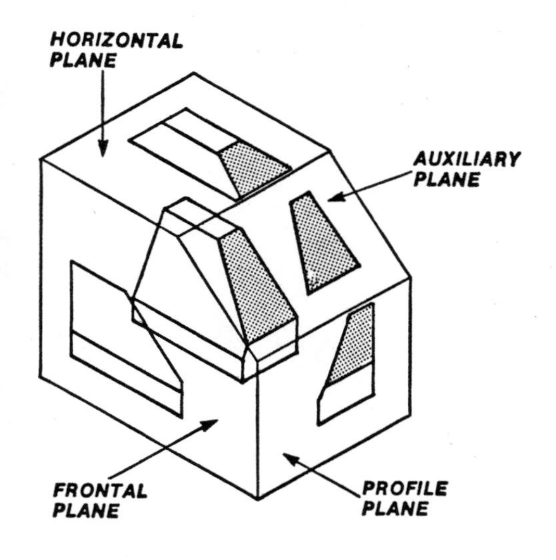

figure 13-2
projection of an inclined surface onto an auxiliary viewing plane

inclined surfaces

The drafter creates an auxiliary view of an inclined surface to show true size and shape. This is accomplished by projecting perpendicular to the inclined surface onto a viewing plane (Figure 13-2).

Usually an auxiliary view shows only the inclined surface. The view is named according to the type of inclined surface (Figure 13-3). Hidden lines are usually omitted in auxiliary views.

Often, when auxiliary views are used, one or more of the principal orthographic views is shown as a partial view (Figure 13-4).

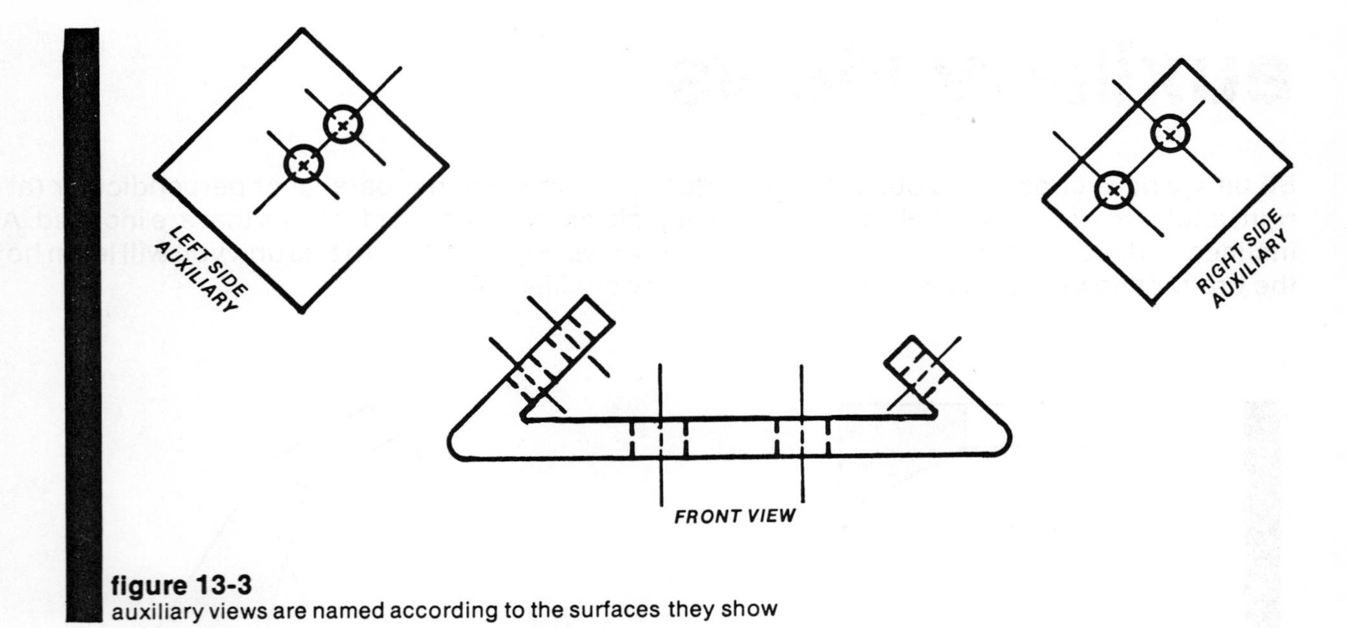

figure 13-3
auxiliary views are named according to the surfaces they show

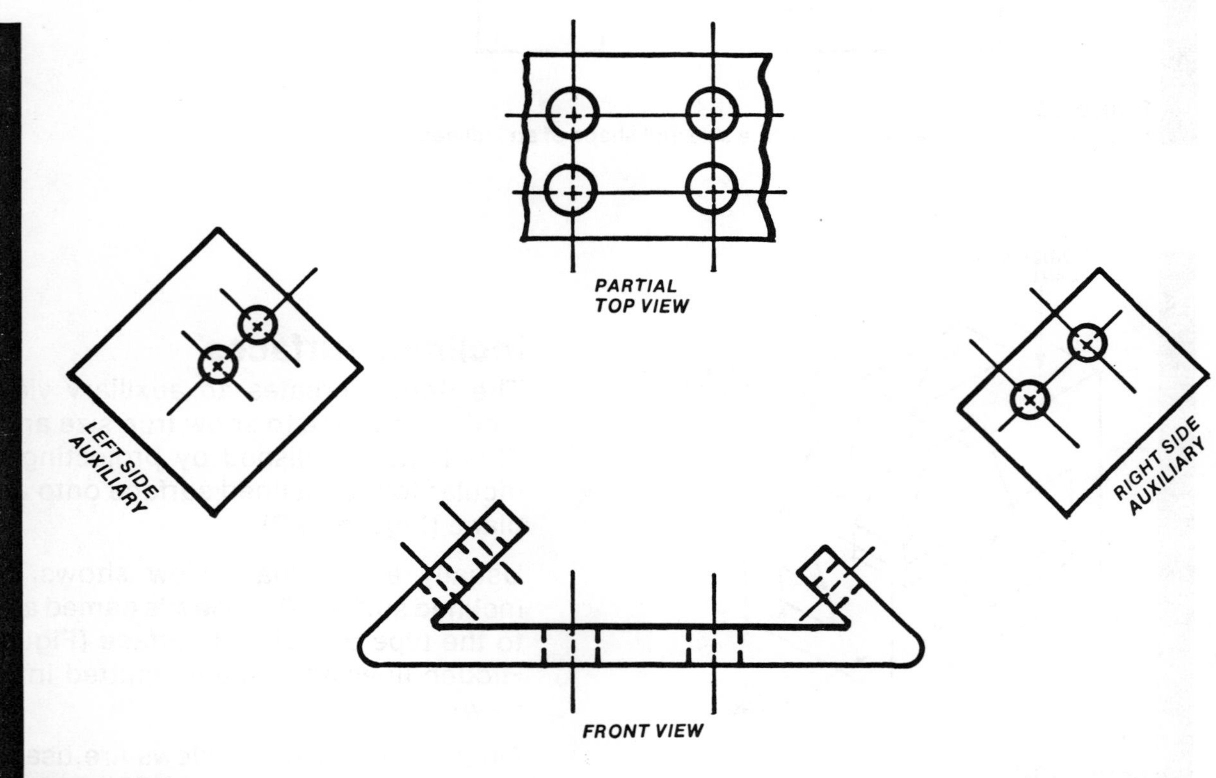

figure 13-4
auxiliary views with a partial main view

self-check quiz 13-a

Identify each view:

Front view ______________________

Right auxiliary ______________________

Partial top view ______________________

Auxiliary elevation ______________________

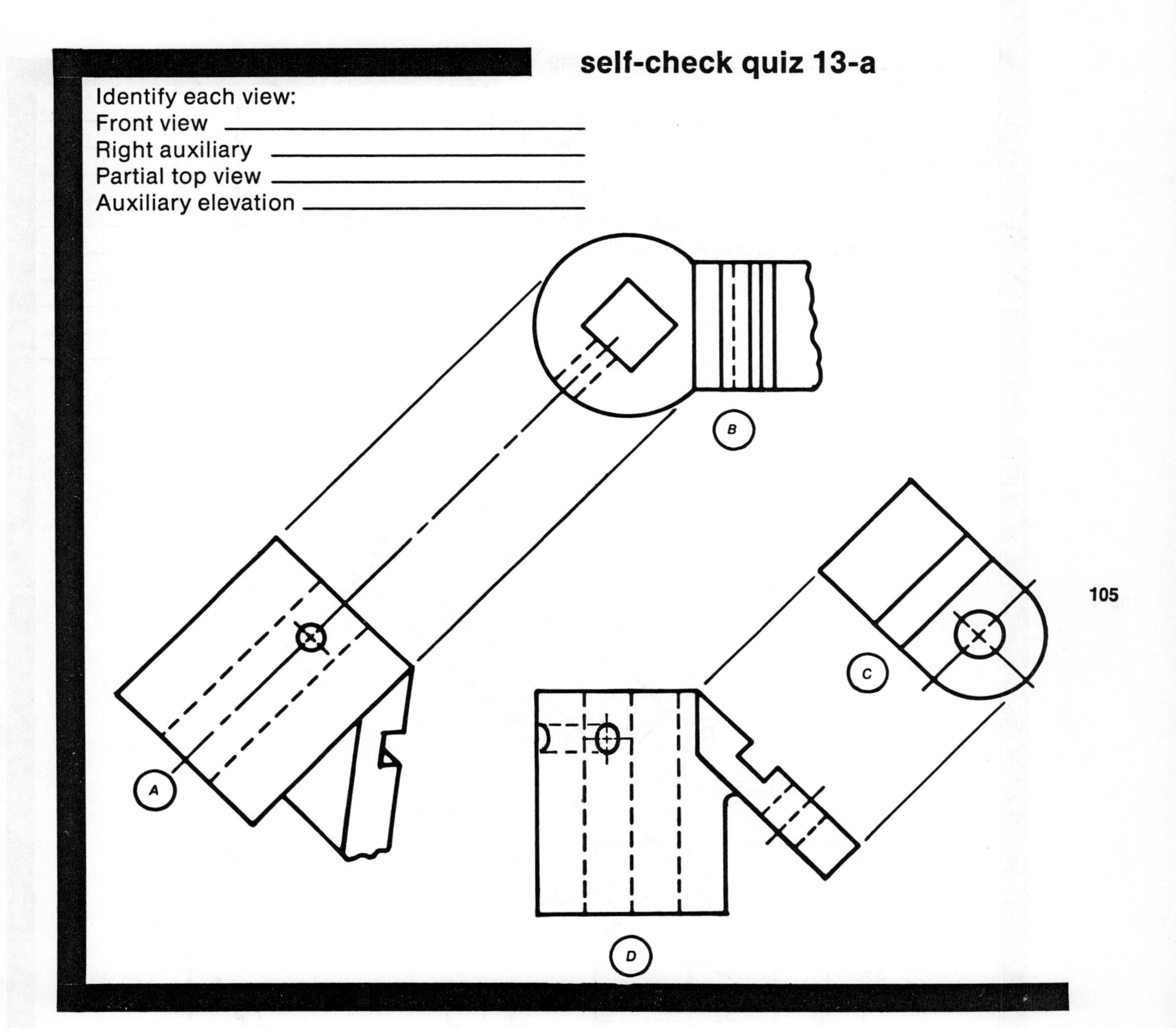

exercise 13-1

______________________________ name

1. Which view shows the ⅜ diameter holes best? 1. ____________
2. Which view shows the slotted hole best? 2. ____________
3. What is the length of the slotted hole? 3. ____________
4. What is the width of the slotted hole? 4. ____________
5. Would a full front view be better than the partial view shown? 5. ____________
6. Determine dimensions A through E. 6. A ____________
B ____________
C ____________
D ____________
E ____________

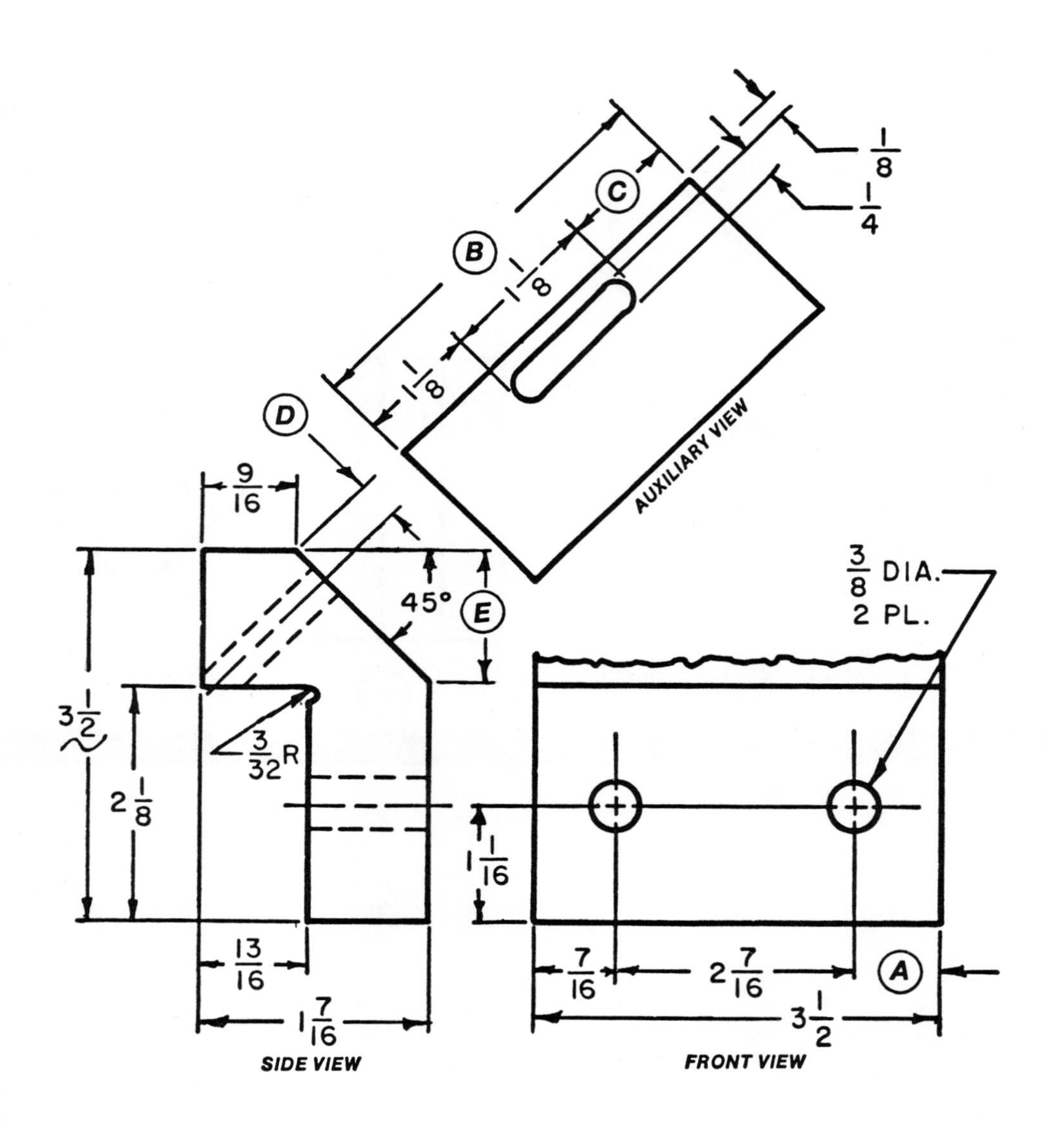

exercise 13-2

______________________________ name

1. Name each view.
2. Identify the following surfaces or lines in an adjacent view:
   A
   B
   C
   D
   E
   F
   G
3. Determine the following dimensions:
   P
   Q
   R
   S
   T
4. What is the upper limit of the 1 1/4 RAD?
5. What is the lower limit of the 9/16 DIA holes?

1. ______
2.
   A. ______
   B. ______
   C. ______
   D. ______
   E. ______
   F. ______
   G. ______
3.
   P. ______
   Q. ______
   R. ______
   S. ______
   T. ______
4. ______
5. ______

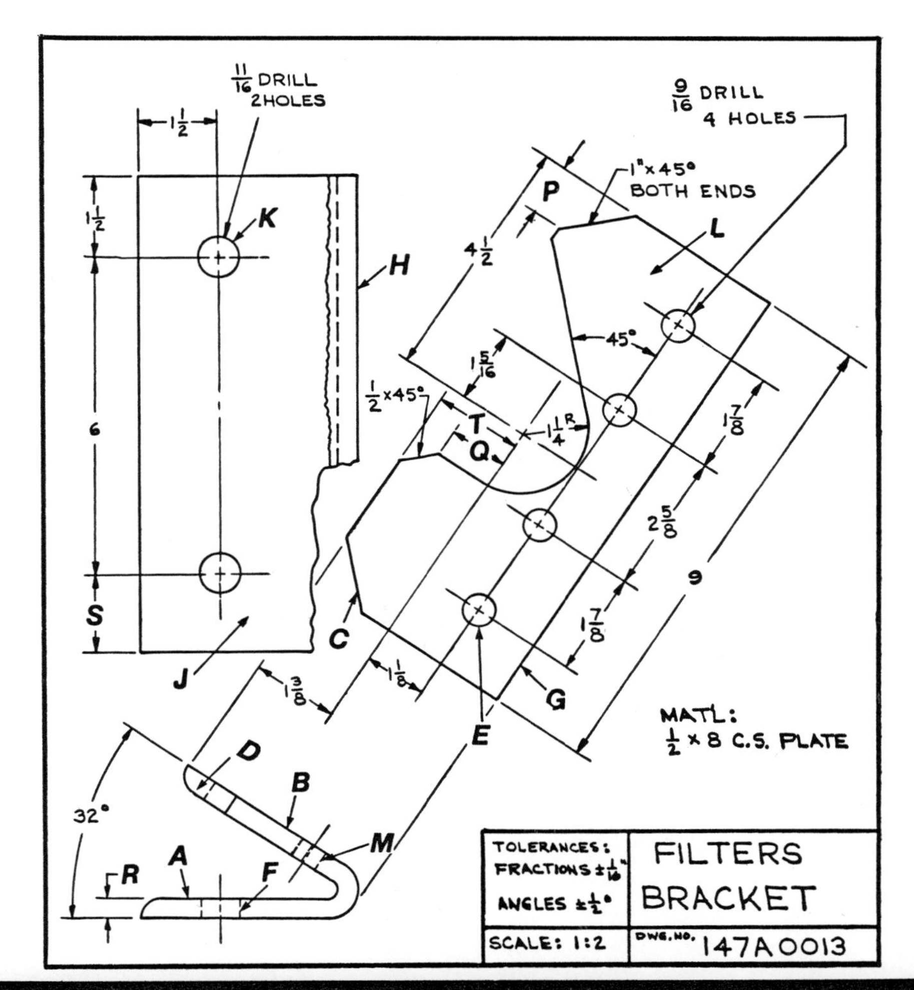

This page for student notes

threads and fasteners

A fastener is a device for joining two or more parts together. Fasteners fall into two categories: permanent (such as rivets or welds) and removable (such as bolts, screws, and pins). After completing this unit you will be able to identify and interpret the different kinds of removable fasteners specified on machine drawings.

threaded fasteners

Threads are a means of fastening parts together. Nuts, bolts, and screws are called *threaded fasteners.*

Drafters show threaded portions of objects in one of three ways: pictorial, schematic, or simplifed (Figure 14-1). *External threads* are threads machined on the outside of an object, such as a bolt. *Internal threads* are threads machined on the inside of an object, such as a nut. Internal threads are also referred to as *tapped threads.* Figure 14-1 shows the tap drill going deeper than the thread depth. The dimension given by the drafter is for the threads and does not include the extra depth of the tap drill. If threads are required clear to the bottom of a hole, a special bottoming tap must be used.

thread forms

Thread forms in common use are based upon industry standards. The most widely used thread form is the Unified Standard Thread (also called Unified Thread, Figure 14-2, *top*).

The older American National Thread (Figure 14-2, *bottom*) is also still used. The Unified and American National are similar. Other thread forms which have special uses other than fastening are square threads, acme threads and buttress threads (Figure 14-3). You will learn how to machine the different thread forms in the machine shop.

thread series

Coarse threads are used for nuts, bolts and screws for general application. Coarse

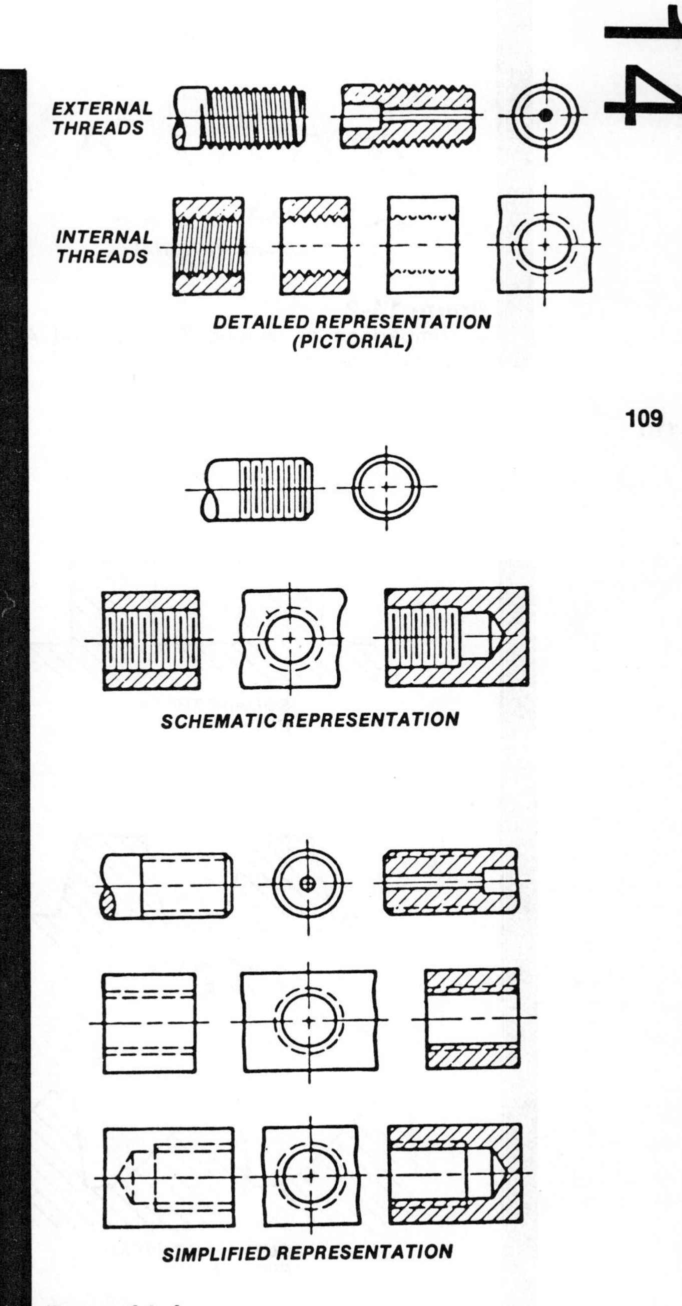

figure 14-1
different conventions for representing screw threads (ANSI Y14.6)

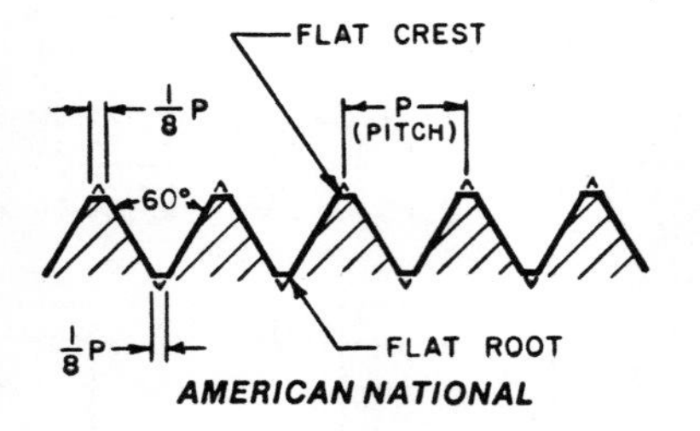

figure 14-2
a comparison of Unified Threads and American National Threads

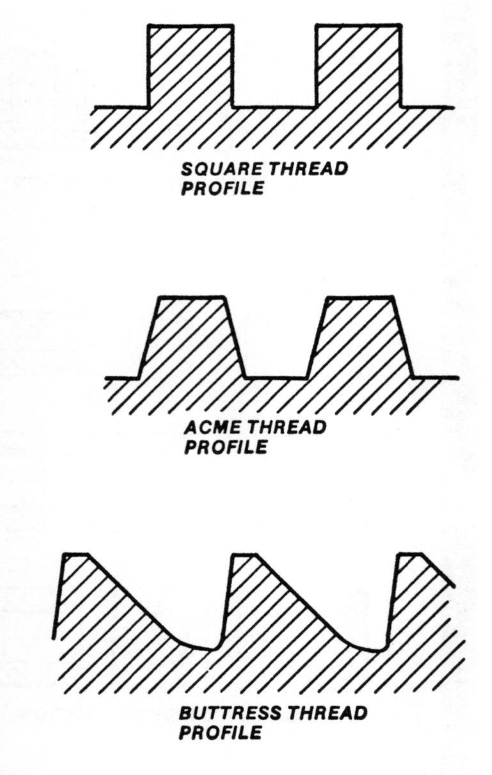

figure 14-3
thread forms with special application such as power transmission

threads are designated UNC (Unified Coarse) or NC (National Coarse).

Fine threads, are used for nuts, bolts and screws where higher strength or vibration resistance is required. They are designated UNF (Unified National Fine) or NF (National Fine). Fine threads are not cut as deeply as coarse threads, for the same basic size, but there are more fine threads in a given length of threaded fastener.

Extra-fine threads are used for special applications such as thin wall tubing, mechanical tubing fasteners and large diameter retaining nuts. They are designated UNEF (Unified Extra Fine) or NEF (Extra Fine).

Constant pitch threads also are for special applications. They come in several series, such as 8, 12 or 16 threads per inch of threaded length. Designations are 8UN, 12UN and 16UN respectively.

Figure 14-4 shows the Unified Thread Series with standard sizes. The first column gives the preferred sizes.

thread class

The class of threads involves the tightness of fit between an externally threaded part and its mating, internally threaded part. *Class 1* applies to threads with a loose fit for ease of quick disassembly. *Class 2* applies to a medium fit for general purpose use. *Class 3* applies to a close fit for applications where close tolerances and minimum play are essential, such as the lead screw of a lathe.

thread designations

The drafter specifies threads by designations such as that in Figure 14-5. The first part of a thread designation is the major (nominal) diameter of the threads in inches. The second part is the number of threads per inch of length. Next is the thread series. After the thread series comes the thread class. Then letter A or B, depending on whether the thread

UNIFIED SCREW THREAD STANDARD SERIES

Nominal Size		Basic Major (Nominal) Diameter	Graded Pitch Series			Constant Pitch Series								Nominal Size
			Coarse UNC	Fine UNF	Extra-Fine UNEF	4 UN	6 UN	8 UN	12 UN	16 UN	20 UN	28 UN	32 UN	
Preferred	Secondary		Threads Per Inch											
0		0.0600	—	80	—	—	—	—	—	—	—	—	—	0
	1	0.0730	64	72	—	—	—	—	—	—	—	—	—	1
2		0.0860	56	64	—	—	—	—	—	—	—	—	—	2
	3	0.0990	48	56	—	—	—	—	—	—	—	—	—	3
4		0.1120	40	48	—	—	—	—	—	—	—	—	—	4
5		0.1250	40	44	—	—	—	—	—	—	—	—	—	5
6		0.1380	32	40	—	—	—	—	—	—	—	—	UNC	6
8		0.1640	32	36	—	—	—	—	—	—	—	—	UNC	8
10		0.1900	24	32	—	—	—	—	—	—	—	—	UNF	10
	12	0.2160	24	28	32	—	—	—	—	—	—	UNF	UNEF	12
1/4		0.2500	20	28	32	—	—	—	—	—	UNC	UNF	UNEF	1/4
5/16		0.3125	18	24	32	—	—	—	—	—	20	28	UNEF	5/16
3/8		0.3750	16	24	32	—	—	—	—	UNC	20	28	UNEF	3/8
7/16		0.4375	14	20	28	—	—	—	—	16	UNF	UNEF	32	7/16
1/2		0.5000	13	20	28	—	—	—	—	16	UNF	UNEF	32	1/2
9/16		0.5625	12	18	24	—	—	—	UNC	16	20	28	32	9/16
5/8		0.6250	11	18	24	—	—	—	12	16	20	28	32	5/8
	11/16	0.6875	—	—	24	—	—	—	12	16	20	28	32	11/16
3/4		0.7500	10	16	20	—	—	—	12	UNF	UNEF	28	32	3/4
	13/16	0.8125	—	—	20	—	—	—	12	16	UNEF	28	32	13/16
7/8		0.8750	9	14	20	—	—	—	12	16	UNEF	28	32	7/8
	15/16	0.9375	—	—	20	—	—	—	12	16	UNEF	28	32	15/16
1		1.0000	8	12	20	—	—	UNC	UNF	16	UNEF	28	32	1
	1-1/16	1.0625	—	—	18	—	—	8	12	16	20	28	—	1-1/16
1-1/8		1.1250	7	12	18	—	—	8	UNF	16	20	28	—	1-1/8
	1-3/16	1.1875	—	—	18	—	—	8	12	16	20	28	—	1-3/16
1-1/4		1.2500	7	12	18	—	—	8	UNF	16	20	28	—	1-1/4
	1-5/16	1.3125	—	—	18	—	—	8	12	16	20	28	—	1-5/16
1-3/8		1.3750	6	12	18	—	UNC	8	UNF	16	20	28	—	1-3/8
	1-7/16	1.4375	—	—	18	—	6	8	12	16	20	28	—	1-7/16
1-1/2		1.5000	6	12	18	—	UNC	8	UNF	16	20	28	—	1-1/2
	1-9/16	1.5625	—	—	18	—	6	8	12	16	20	—	—	1-9/16
1-5/8		1.6250	—	—	18	—	6	8	12	16	20	—	—	1-5/8
	1-11/16	1.6875	—	—	18	—	6	8	12	16	20	—	—	1-11/16
1-3/4		1.7500	5	—	—	—	6	8	12	16	20	—	—	1-3/4
	1-13/16	1.8125	—	—	—	—	6	8	12	16	20	—	—	1-13/16
1-7/8		1.8750	—	—	—	—	6	8	12	16	20	—	—	1-7/8
	1-15/16	1.9375	—	—	—	—	6	8	12	16	20	—	—	1-15/16
2		2.0000	4½	—	—	—	6	8	12	16	20	—	—	2
	2-1/8	2.1250	—	—	—	—	6	8	12	16	20	—	—	2-1/8
2-1/4		2.2500	4½	—	—	—	6	8	12	16	20	—	—	2-1/4
	2-3/8	2.3750	—	—	—	—	6	8	12	16	20	—	—	2-3/8
2-1/2		2.5000	4	—	—	UNC	6	8	12	16	20	—	—	2-1/2
	2-5/8	2.6250	—	—	—	4	6	8	12	16	20	—	—	2-5/8
2-3/4		2.7500	4	—	—	UNC	6	8	12	16	20	—	—	2-3/4
	2-7/8	2.8750	—	—	—	4	6	8	12	16	20	—	—	2-7/8
3		3.0000	4	—	—	UNC	6	8	12	16	20	—	—	3
	3-1/8	3.1250	—	—	—	4	6	8	12	16	—	—	—	3-1/8
3-1/4		3.2500	4	—	—	UNC	6	8	12	16	—	—	—	3-1/4
	3-3/8	3.3750	—	—	—	4	6	8	12	16	—	—	—	3-3/8
3-1/2		3.5000	4	—	—	UNC	6	8	12	16	—	—	—	3-1/2
	3-5/8	3.6250	—	—	—	4	6	8	12	16	—	—	—	3-5/8
3-3/4		3.7500	4	—	—	UNC	6	8	12	16	—	—	—	3-3/4
	3-7/8	3.8750	—	—	—	4	6	8	12	16	—	—	—	3-7/8
4		4.0000	4	—	—	UNC	6	8	12	16	—	—	—	4
	4-1/8	4.1250	—	—	—	4	6	8	12	16	—	—	—	4-1/8
4-1/4		4.2500	—	—	—	4	6	8	12	16	—	—	—	4-1/4
	4-3/8	4.3750	—	—	—	4	6	8	12	16	—	—	—	4-3/8
4-1/2		4.5000	—	—	—	4	6	8	12	16	—	—	—	4-1/2
	4-5/8	4.6250	—	—	—	4	6	8	12	16	—	—	—	4-5/8
4-3/4		4.7500	—	—	—	4	6	8	12	16	—	—	—	4-3/4
	4-7/8	4.8750	—	—	—	4	6	8	12	16	—	—	—	4-7/8
5		5.0000	—	—	—	4	6	8	12	16	—	—	—	5
	5-1/8	5.1250	—	—	—	4	6	8	12	16	—	—	—	5-1/8
5-1/4		5.2500	—	—	—	4	6	8	12	16	—	—	—	5-1/4
	5-3/8	5.3750	—	—	—	4	6	8	12	16	—	—	—	5-3/8
5-1/2		5.5000	—	—	—	4	6	8	12	16	—	—	—	5-1/2
	5-5/8	5.6250	—	—	—	4	6	8	12	16	—	—	—	5-5/8
5-3/4		5.7500	—	—	—	4	6	8	12	16	—	—	—	5-3/4
	5-7/8	5.8750	—	—	—	4	6	8	12	16	—	—	—	5-7/8
6		6.0000	—	—	—	4	6	8	12	16	—	—	—	6

figure 14-4
a chart showing the unified thread series. The column to the far left shows the preferred sizes

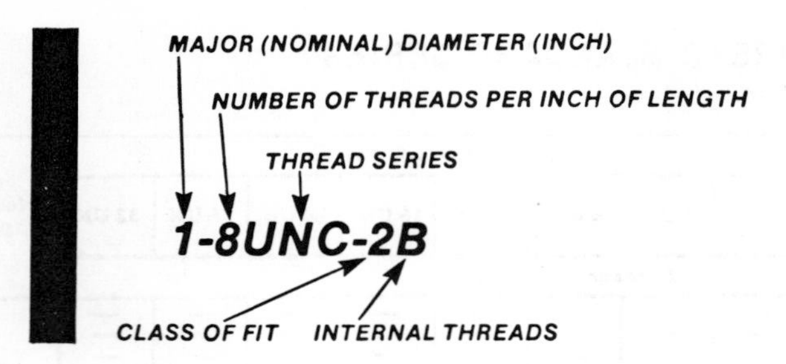

figure 14-5
a thread designation

3/4-10UNC-2A-LH

figure 14-6
designation of a left-hand thread. This is an external (A) left-hand (LH) thread with 10 unified national coarse threads per inch on a 3/4 (0.7500) inch diameter.

is external (A) or internal (B). Finally, additional information may be added at the end of the thread designation, such as the length of the threaded portion, or other machining information.

Threads are either right-hand or left-hand. Right-hand threads are most common. A part with right-hand threads advances, or is tightened, by turning to the right (clockwise). A part with left-hand threads advances, or is tightened, by turning to the left (counterclockwise). If a thread is right-hand, no notation in the thread designation is used. If a thread is left-hand, the notation LH is added at the end of the thread designation (Figure 14-6).

self-check quiz 14-a

For the object shown, identify the following:

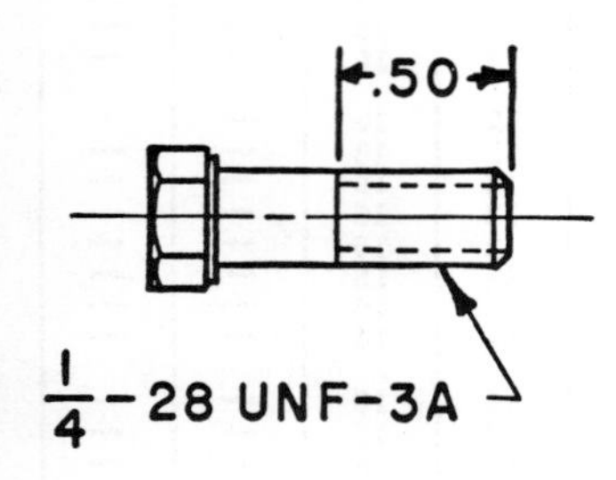

1. Type of thread representation ______________
2. Thread form ______________
3. Thread class ______________
4. Internal or external threads ______________
5. Number of threads per inch ______________
6. Total number of threads ______________
7. Thread series ______________
8. Thread diameter (size) ______________

pipe threads

There are two forms of threads used to join pipe, tapered and straight. Tapered pipe threads are designated by the nominal pipe size and the letters NPT (Figure 14-7). Tapered threads are tapered 3/16″ per inch of thread. Straight pipe threads are designated by the nominal pipe size and the letters NPS (Figure 14-8).

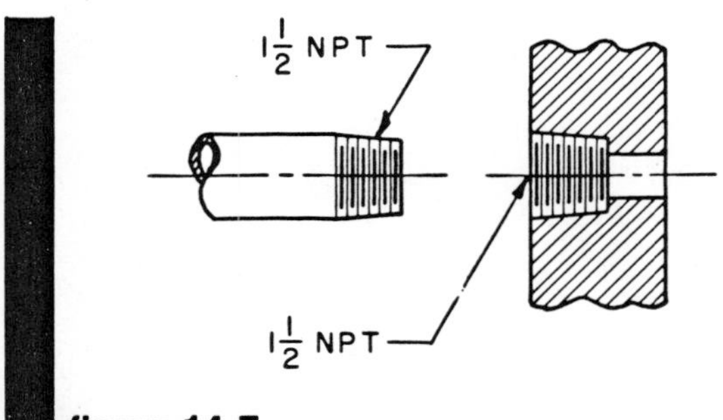

figure 14-7
tapered pipe threads and their designation

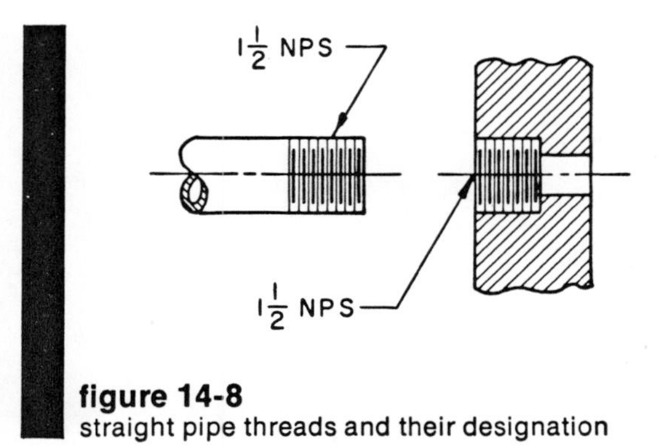

figure 14-8
straight pipe threads and their designation

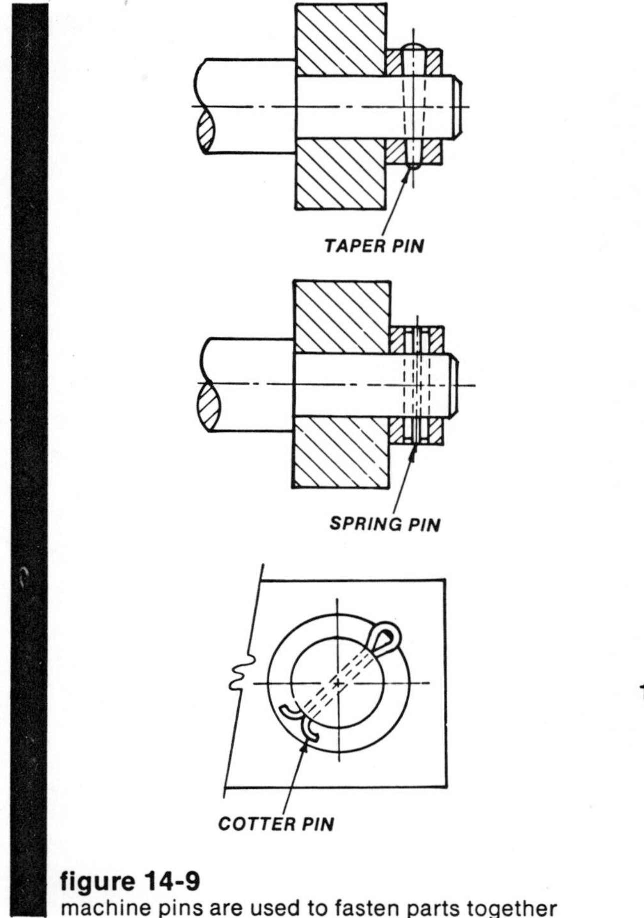

figure 14-9
machine pins are used to fasten parts together

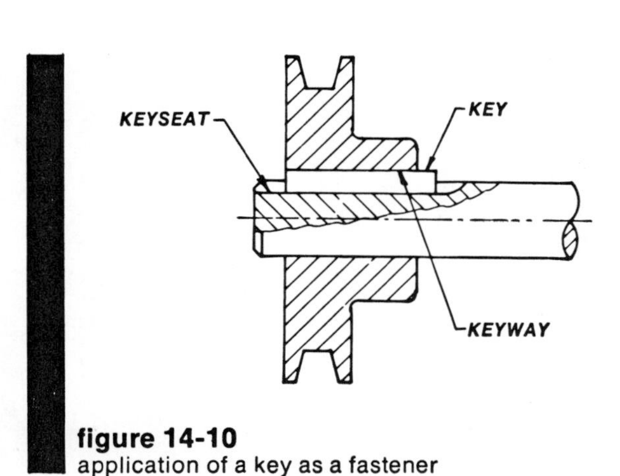

figure 14-10
application of a key as a fastener

pins and keys

Parts may be fastened together by pins. Figure 14-9 shows several different kinds of pins. Mating parts must have holes for the pins. Dowel pins require close fitting holes to keep the mating parts in good alignment.

Keys are used to fasten pulleys, gears and other parts to shafts. Keys may be square, flat, tapered or rounded. The machined surface on a shaft where the key fits is called a *keyseat.* The machined surface on the mating hub is called a *keyway.* Figure 14-10 shows an application of a key as a fastener. Keyways and keyseats may be detailed with dimensions, or they may be defined by a local note such as "1/4 X 1/8 KEYWAY 2 1/4 LONG."

This page for student notes

exercise 14-1

_______________________________ name

1. Determine the following dimensions:
   A
   B
   C
   D
   E
   F
2. What does "UNC" stand for in the .38 tapped thread designation?
3. How many total threads are there for the .75 threaded portion?
4. What class fit is required for the .75 threaded portion?
5. What does the letter "A" mean in the thread designation?
6. Which diameters must be machined concentric?
7. A number 28 drill is to be used to drill a hole through the motor shaft extension. What is the hole for?

1.
A ______
B ______
C ______
D ______
E ______
F ______
2. ______
3. ______
4. ______
5. ______
6. ______
7. ______

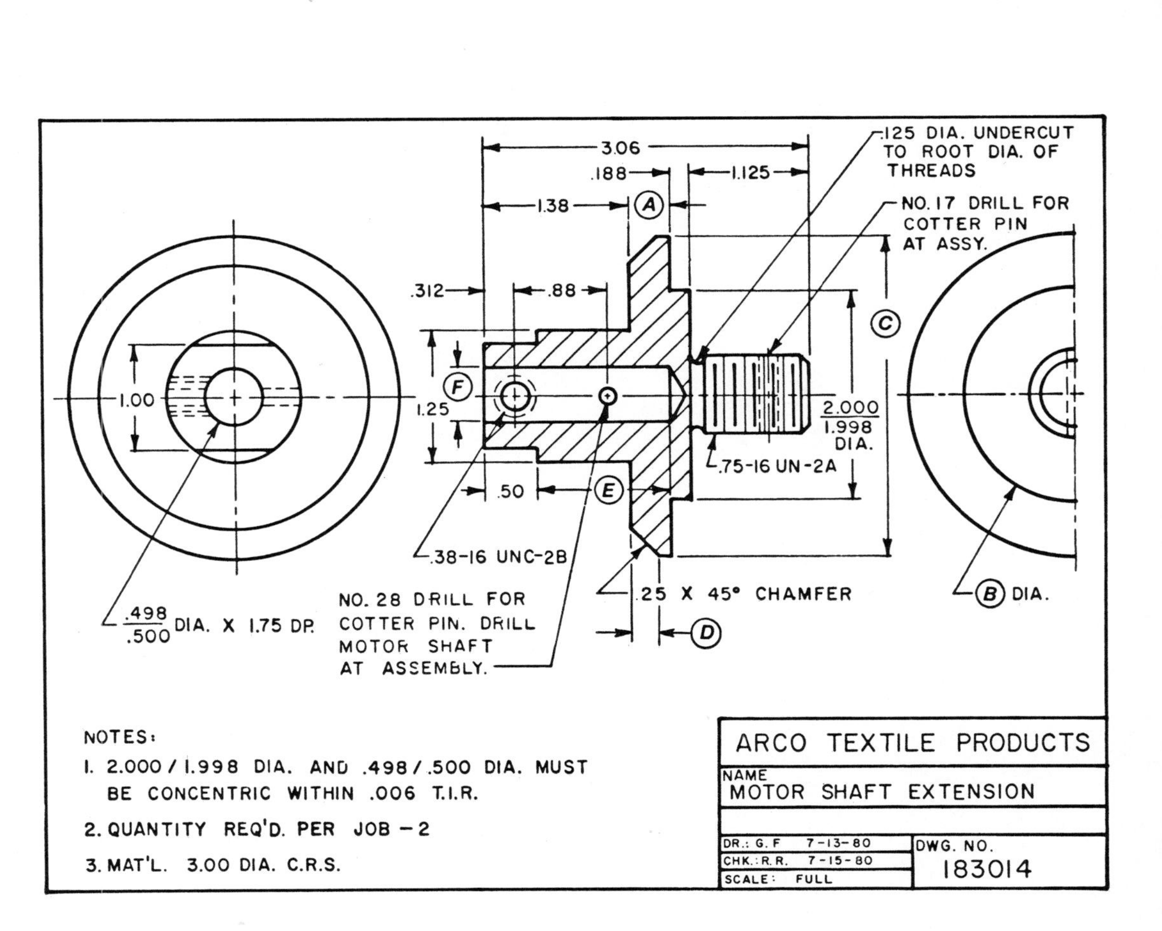

This page for student notes

surface finish

Many different processes are used to machine parts. Each process such as planing, drilling, reaming or honing has a characteristic surface finish associated with that process. Surface finish includes roughness, waviness and lay. The drafter places finish marks on the drawing to define the required surface finish. In this unit you will learn how to identify the surface finish requirements of a drawing.

finish designation

The most common finish requirement on a drawing is surface roughness. Roughness is measured in microinches (μ in.) or millionths of an inch average deviation from the mean surface (Figure 15-1).

In addition to surface roughness, it is sometimes desirable to control the waviness of a surface. Waviness results from vibration, deflection, heat treatment and working strains. It is measured in thousandths of an inch (Figure 15-2).

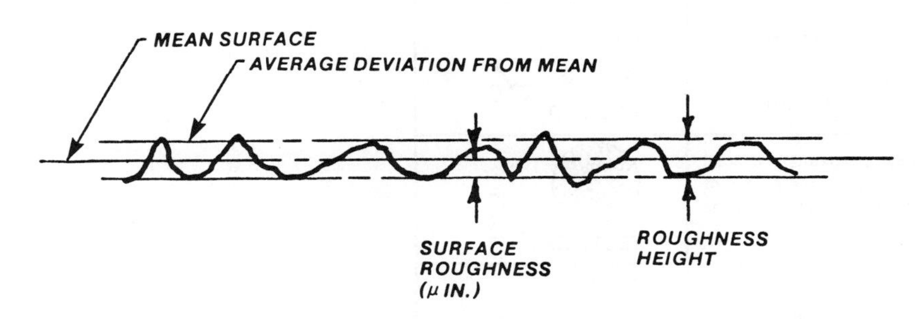

figure 15-1
a magnified view of the surface of an object showing how surface roughness is defined

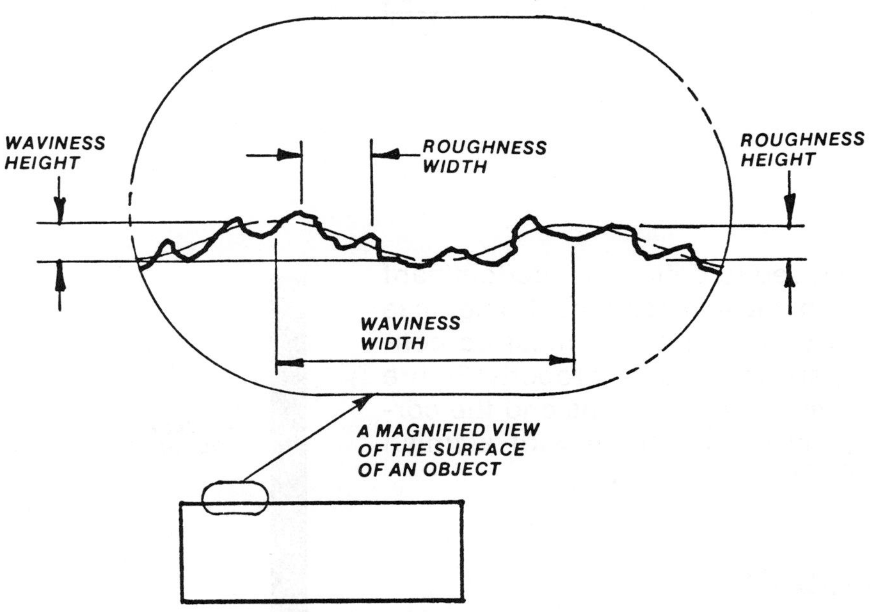

figure 15-2
roughness and waviness characteristics

Lay Symbol	Meaning	Example Showing Direction of Tool Marks
=	Lay approximately parallel to the line representing the surface to which the symbol is applied.	
⊥	Lay approximately perpendicular to the line representing the surface to which the symbol is applied.	
X	Lay angular in both directions to line representing the surface to which the symbol is applied.	
M	Lay multidirectional.	
C	Lay approximately circular relative to the center of the surface to which the symbol is applied.	
R	Lay approximately radial relative to the center of the surface to which the symbol is applied.	
P[3]	Lay particulate, non-directional, or protuberant.	

figure 15-3
lay symbols
(ANSI 14.36, 1978)

Lay is a term used to define the predominant surface pattern due to machining. Sometimes the direction of lay on a part must be controlled if the part is to function properly. Figure 15-3 shows several lay patterns and the corresponding lay designations with a finish symbol (√‾‾).

finish symbols

Figure 15-4 shows the standard finish symbol with appropriate finish designations. The symbol is shown on an edge view of the sur-

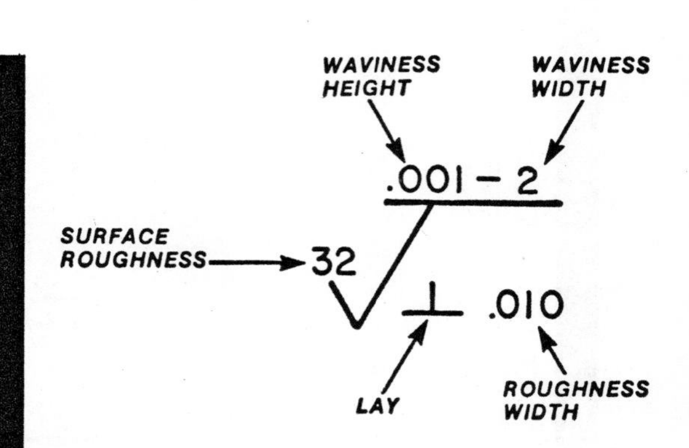

figure 15-4
surface finish symbol. Surface roughness is in millionths (μ) of an inch.

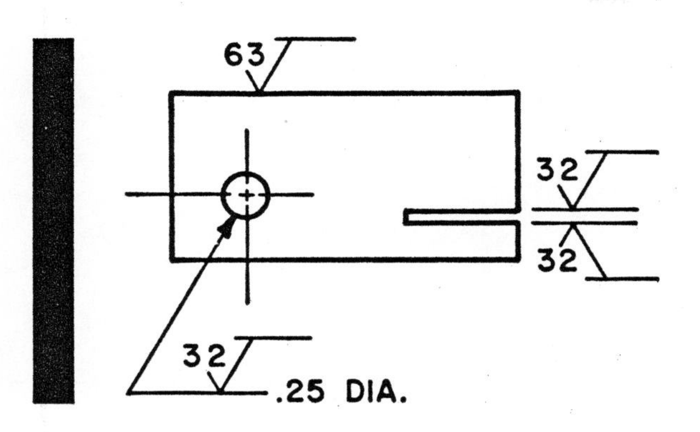

figure 15-5
application of surface finish symbols

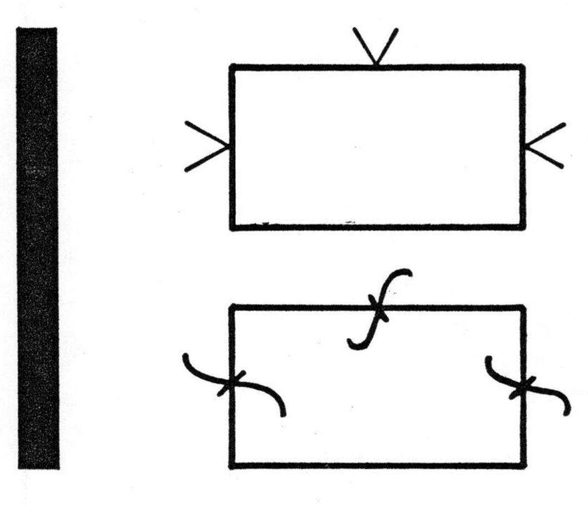

figure 15-6
old surface finish symbols, rarely used today

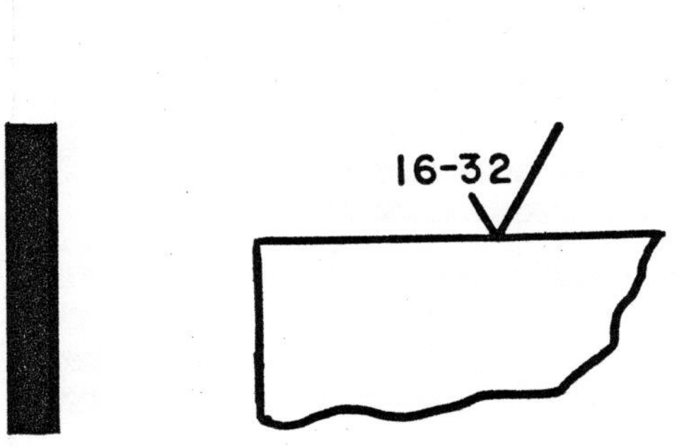

figure 15-7
specifying maximum and minimum values for surface roughness

face to which it applies. If there is no room on the surface to show the finish symbol, it may be placed on an extension line or leader from the surface. See Figure 15-5.

Sometimes only the surface roughness is specified if the waviness and lay are not critical. Sometimes surface finish is handled by a general note such as "FINISH ALL OVER" or "F.A.O.". Older drawings may show one of the former finish symbols such as shown in Figure 15-6. If the surface is not allowed to be too smooth, upper and lower limits for the surface roughness are specified (Figure 15-7).

Figure 15-8 shows typical surface finishes attainable by various shop processes. Note that higher or lower values may be obtained, depending on conditions.

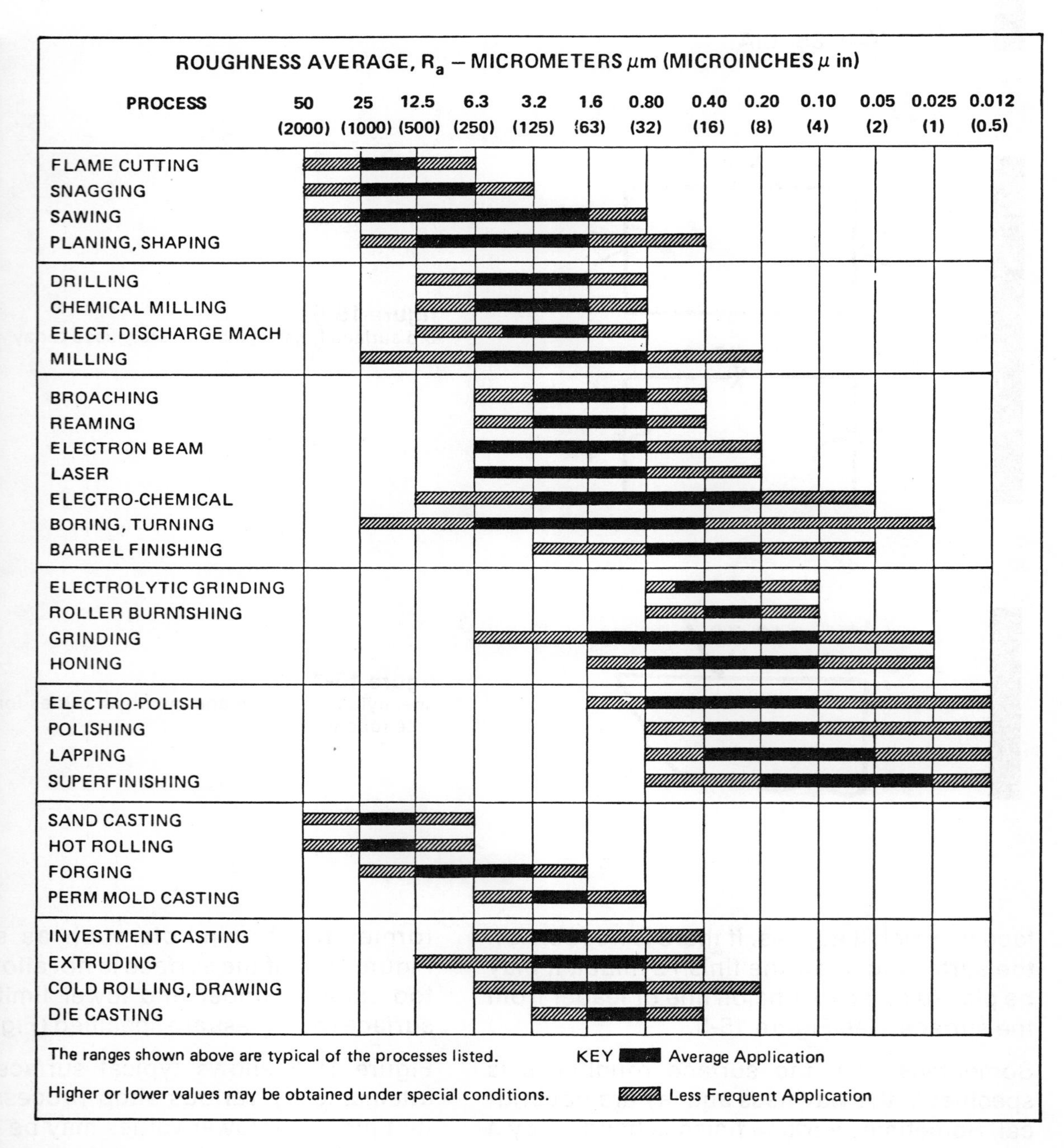

figure 15-8
surface roughness ranges for various shop processes (ANSI B46.1)

exercise 15-1

____________________ name

1. List the surface finish required at surfaces A through L.

1. A ______
   B ______
   C ______
   D ______
   E ______
   F ______
   G ______
   H ______
   I ______
   J ______
   K ______
   L ______

2. What is the permissible waviness height on surface I?
3. What is the required lay of surface I?
4. How many .375 diameter holes are required?
5. Determine dimensions M through R.

2. ______
3. ______
4. ______
5. M ______
   N ______
   P ______
   Q ______
   R ______

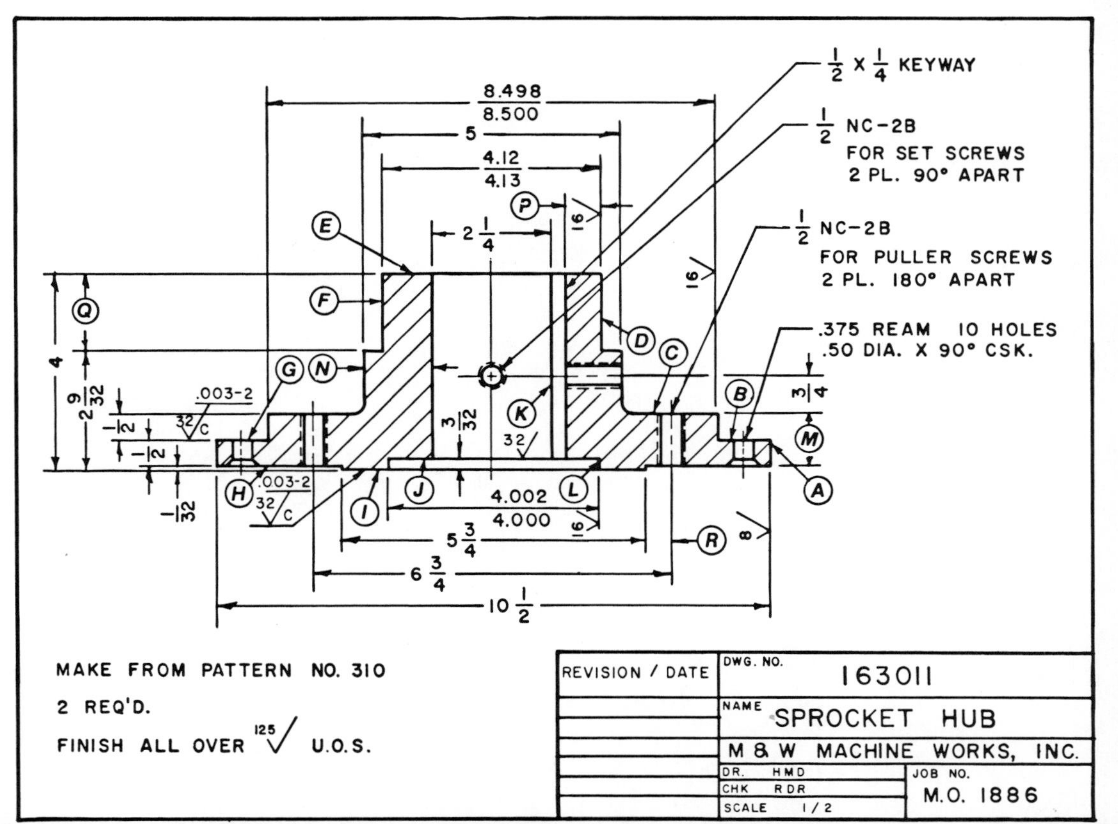

This page for student notes

conventional drafting practices

Unit 16

Drafters take many shortcuts to save time. These shortcuts are known as *conventional drafting practices*. In this unit you will learn to recognize and interpret conventional drafting practices such as multi-part drawings, omissions of standard hardware and omission of repetitive detail.

multi-part drawings and assembly drawings

The drafter sometimes draws more than one part on a single drawing sheet. This is especially true if the parts are to be assembled together. Figure 16-1 shows two parts detailed on the same drawing. There is sufficient information on this drawing to machine both parts.

Figure 16-2 shows three parts in assembly. The purpose of an *assembly drawing* is to show how parts fit together. Usually an

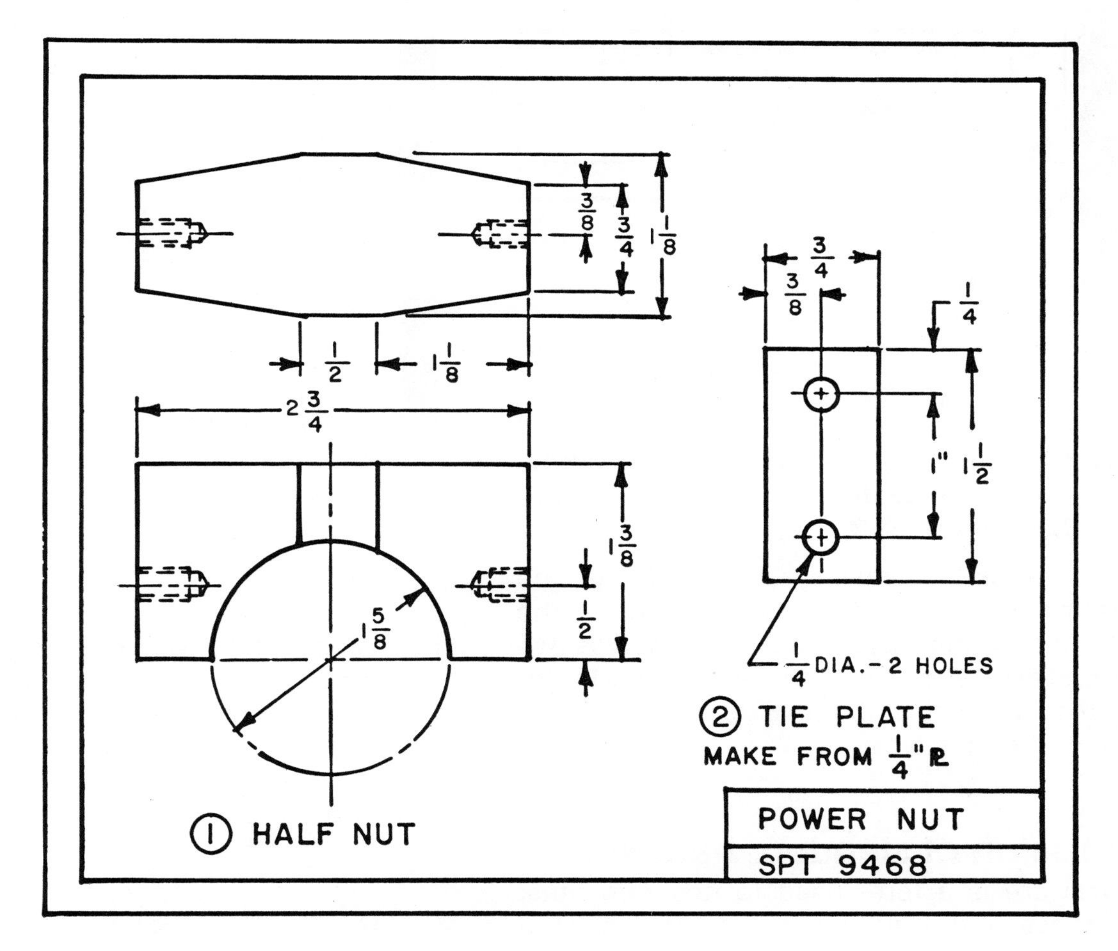

figure 16-1
a multi-part drawing containing two objects

figure 16-2
a typical assembly detail

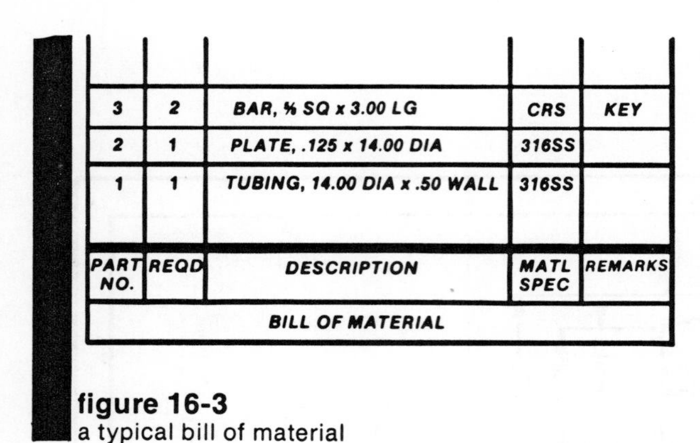

PART NO.	REQD	DESCRIPTION	MATL SPEC	REMARKS
3	2	BAR, ⅜ SQ x 3.00 LG	CRS	KEY
2	1	PLATE, .125 x 14.00 DIA	316SS	
1	1	TUBING, 14.00 DIA x .50 WALL	316SS	
BILL OF MATERIAL				

figure 16-3
a typical bill of material

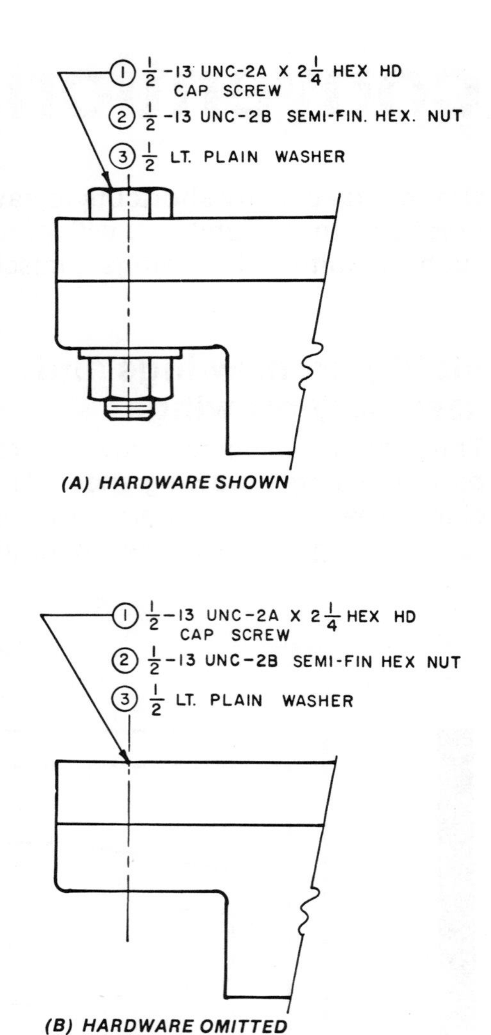

figure 16-4
alternate methods of representing standard hardware

assembly drawing has each part numbered. Information on each part is given in a *Bill of Material* (Figure 16-3).

If the assembly drawing is fairly simple, the drafter puts machining information on the drawing. In this case, the drawing serves two purposes. It shows how to make the parts and how to assemble them. Exercise 16-2 is a drawing of this type.

omission of standard hardware

Hardware is a term used to describe nuts, bolts, washers, pins and other fasteners. If the hardware is a standard size, the drafter may omit showing it on the drawing. Figure 16-4A shows an assembly with nuts and bolts. Figure 16-4B shows the same assembly with only a centerline where the nuts and bolts belong. Any hardware may be shown in this manner, as long as the parts are identified by a part number or note.

omission of repetitive detail

Often the drafter will show a representative detail and indicate that the detail applies in

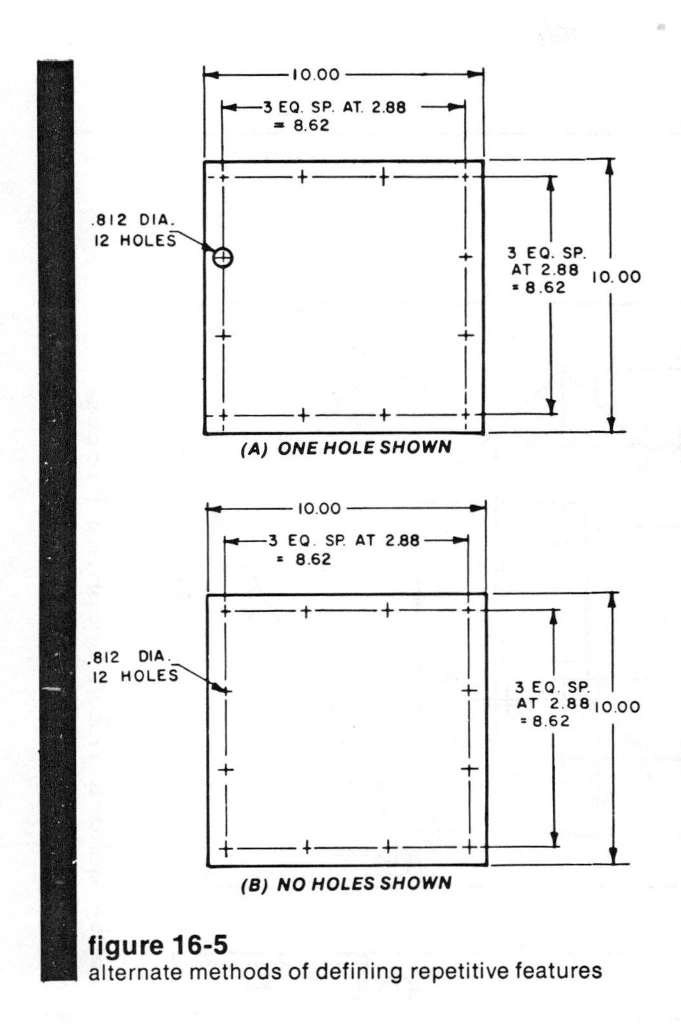

figure 16-5
alternate methods of defining repetitive features

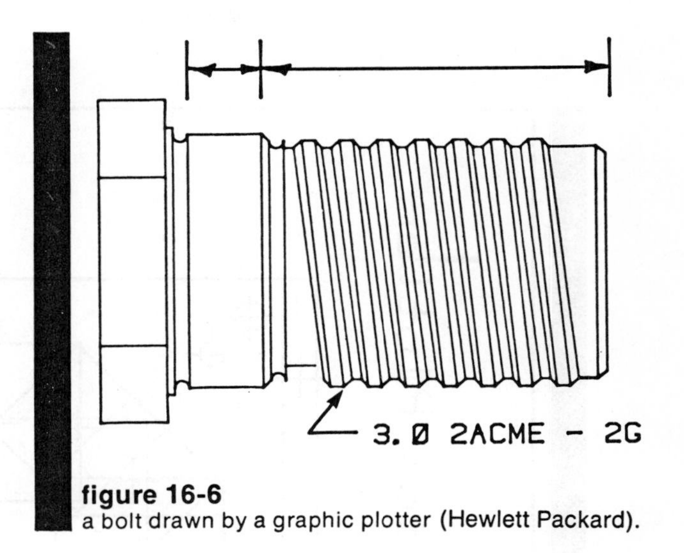

figure 16-6
a bolt drawn by a graphic plotter (Hewlett Packard).

more than one place. Figure 16-5A shows a cover plate with twelve holes required, all the same size. The drafter showed one hole and indicated the rest with centerlines. Sometimes the drafter shows no detail at all, and gives the necessary machining information by a leader. (Figure 16-5B).

computer graphics and numerical control

For the past forty years, computers have helped us to accomplish great things, including sending men to the moon. Computers have also been put to good use in both the drafting room and the machine shop. In the drafting room, a design concept is programmed into a computer. The computer is connected to equipment to produce a drawing. This is known as *computer graphics*. The equipment which draws the lines on paper is called a *graphics plotter.*

We read prints of these drawings just like any other prints. The only difference is that computer graphics drawings appear more "mechanical" since they are machine-produced. Figure 16-6 shows a computer graphics drawing of a bolt.

Computers are also used to control machine tools for making parts in the shop. These machine tools are known as numerically controlled machine tools. Computer graphics is often used to check the machining path of a numerically controlled machine. In this manner, you can check whether or not the machine tool will cut the correct pattern or contour. Figure 16-7 shows the machining desired for a part on the left and what the numerically controlled machine is programmed to do on the right.

Computers have become great assistants to both the drafter and the machinist. You will undoubtedly be involved with computers in the shop.

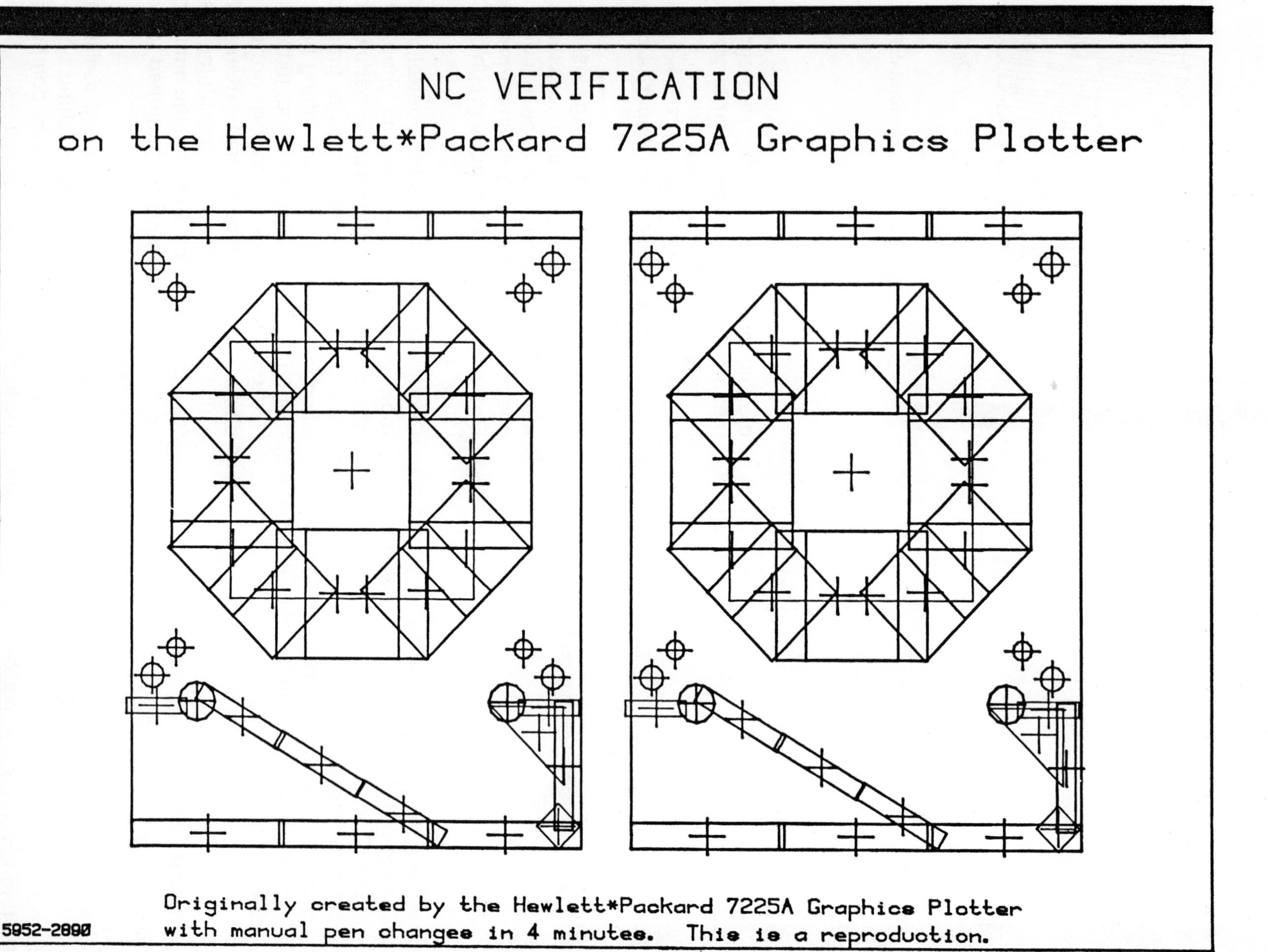

figure 16-7
computer graphics is used to verify the machine contour for a numerical control operation

exercise 16-1

______________________ name

1. How many parts are shown? 1. ________
2. What was the original length of the truck wheel? 2. ________
3. What is the pattern number for the truck wheel? 3. ________
4. What is the material of the truck wheel? 4. ________
5. What is the maximum diameter of the wheel stud? 5. ________
6. What is the minimum bore of the truck wheel? 6. ________
7. What is the nominal width of the retaining ring grooves on the wheel stud? 7. ________
8. What surface finish is required in the retaining ring grooves of the wheel stud? 8. ________
9. What surface finish is required in the retaining ring grooves of the truck wheel? 9. ________
10. Determine dimension A. 10. ________

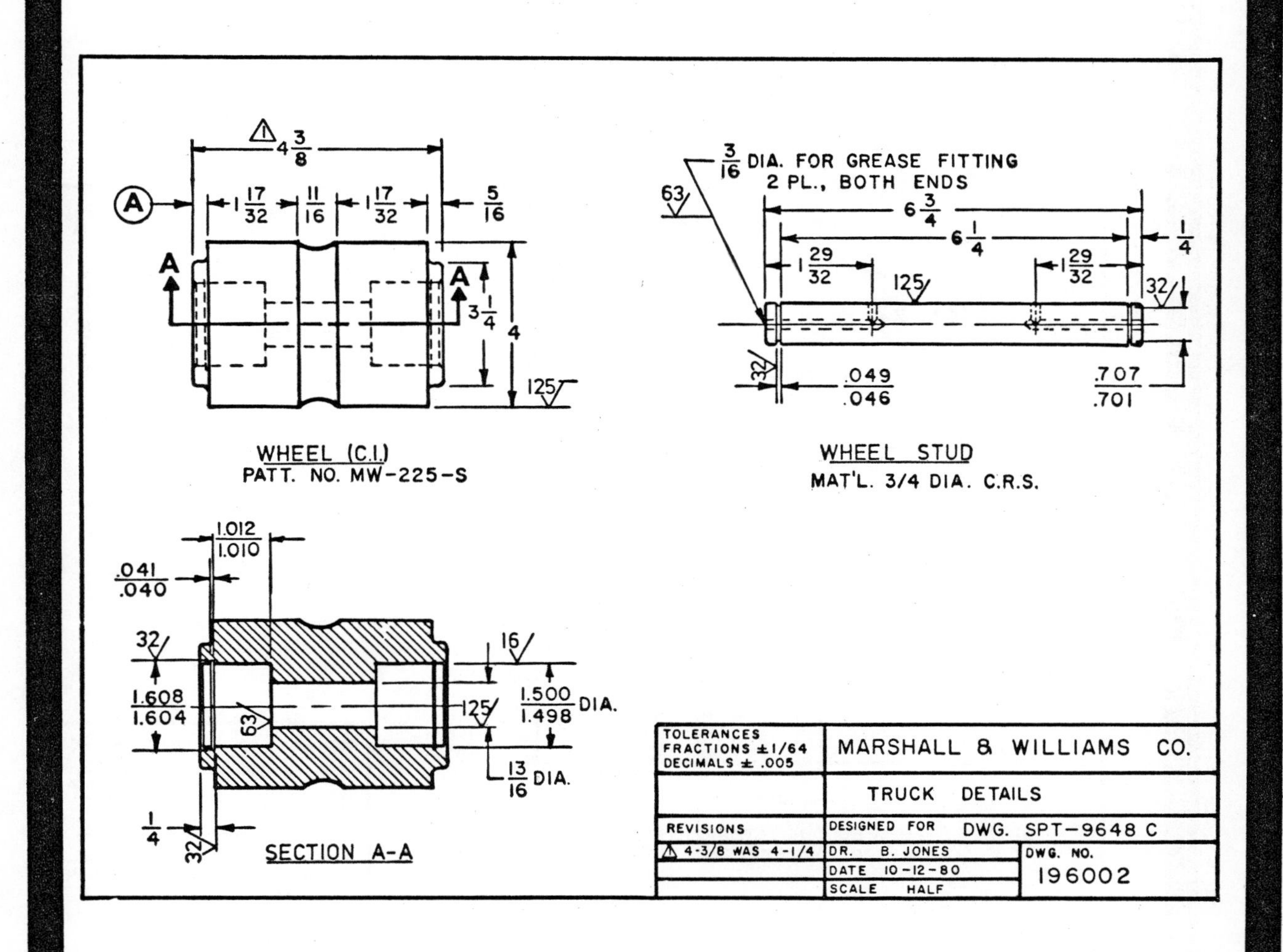

exercise 16-2

______________________ name

1. Is this drawing a machining drawing an assembly drawing or both? 1. ________
2. What is the total number of parts shown on the drawing? 2. ________
3. What size screws are required? 3. ________
4. How long are the screws? 4. ________
5. If the screw holes on a tie plate were machined 1.062 apart, would the tie plate be acceptable? 5. ________
6. What is the nominal diameter of the acme threads? 6. ________
7. What is the required surface finish of the mating surfaces of the split nut? 7. ________
8. Determine dimensions A through D 8. A ________ B ________ C ________ D ________
9. What is the casting drawing number? 9. ________
10. What special procedure must be followed before machining the acme thread? 10. ________
11. Sketch a side view of the assembly in the space provided. 11.

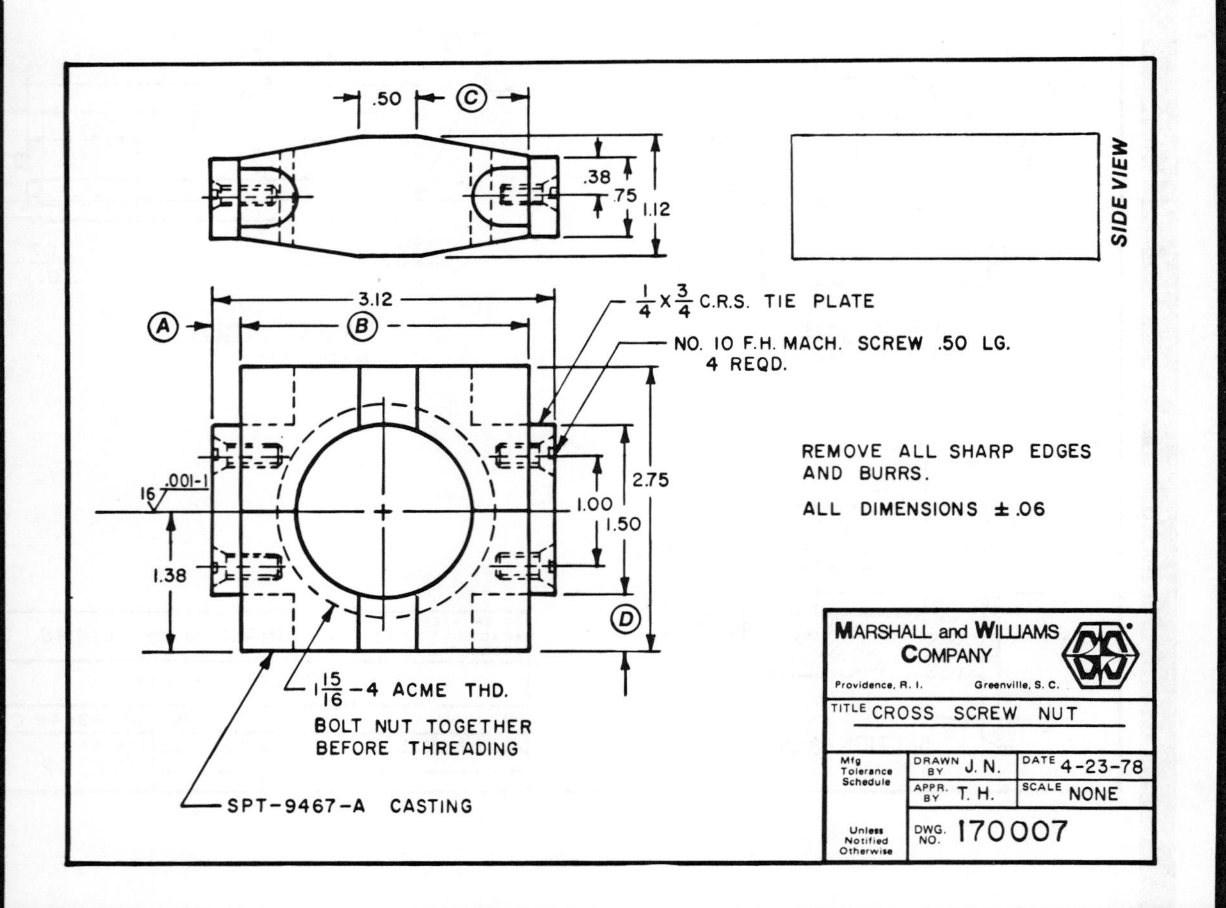

metrics

Unit 17

The metric system of measurement is simpler than our present inch-foot-yard American Customary system. It is easy to learn. Most of the world already uses the metric system. Many major companies such as Coca Cola, Ford, General Electric and Sears Roebuck are already using metrics in their products. We will have to adopt the metric system if we want to be competitive in the world marketplace. After completing this unit you will be able to read metric prints.

The metric system has many advantages. It is simple. Units are understandable and interchangeable. The greatest economic advantage is that parts will be standardized and interchangeable world-wide. The old metric system used for so many years in Europe has been updated to meet the demands of present-day industry. This revised metric system is called *SI*, which stands for *System International d'Unites* (International System of Units).

the basic SI units

The SI system involved six basic units of measurement:

- length
- time
- mass
- temperature
- electric current
- luminous intensity.

For print-reading purposes, we will consider the unit of length only.

MULTIPLES	PREFIX	SYMBOL	EXAMPLE:
1 000 000	mega	M	1Mm = 1000 000m
1000	kilo	k	1km = 1000m
100	hecto	h	1hm = 100m
10	deka	da	1dam = 10m
BASE UNIT			1m = 1m
0.1	deci	d	1dm = 0.1m
0.01	centi	c	1cm = 0.01m
0.001	milli	m	1mm = 0.001m
0.000 001	micro	μ	1μm = 0.000 001m

figure 17-2
prefixes used in the metric system. These prefixes apply to metric units of length, area, volume and weight

figure 17-1
the meter and the yard compared

The basic SI unit of length is the meter (m). A meter is slightly longer than a yard (Figure 17-1). Prefixes (Figure 17-2) are added to the base unit (meter) to create larger or smaller units. Thus 10 meters make one dekameter (dam), 100 meters make one hectometer (hm), and so on. Likewise, 0.1 meter is a decimeter (dm), 0.01 meter is a centimeter (cm), 0.001 meter is a millimeter, and so on.

This *base-10* system is easier to use than inches, feet, yards and miles. Figure 17-3 compares our customary units to metric units.

For drawings, the recommended units for length are millimeters, meters, and kilometers. Drafters avoid expressing length in centimeters, decimeters, dekameters and hecto-

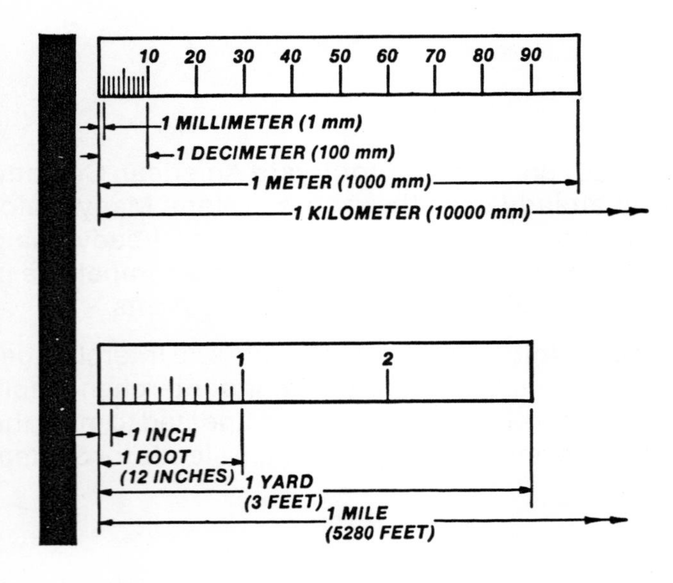

figure 17-3
a comparison of metric and customary units of length

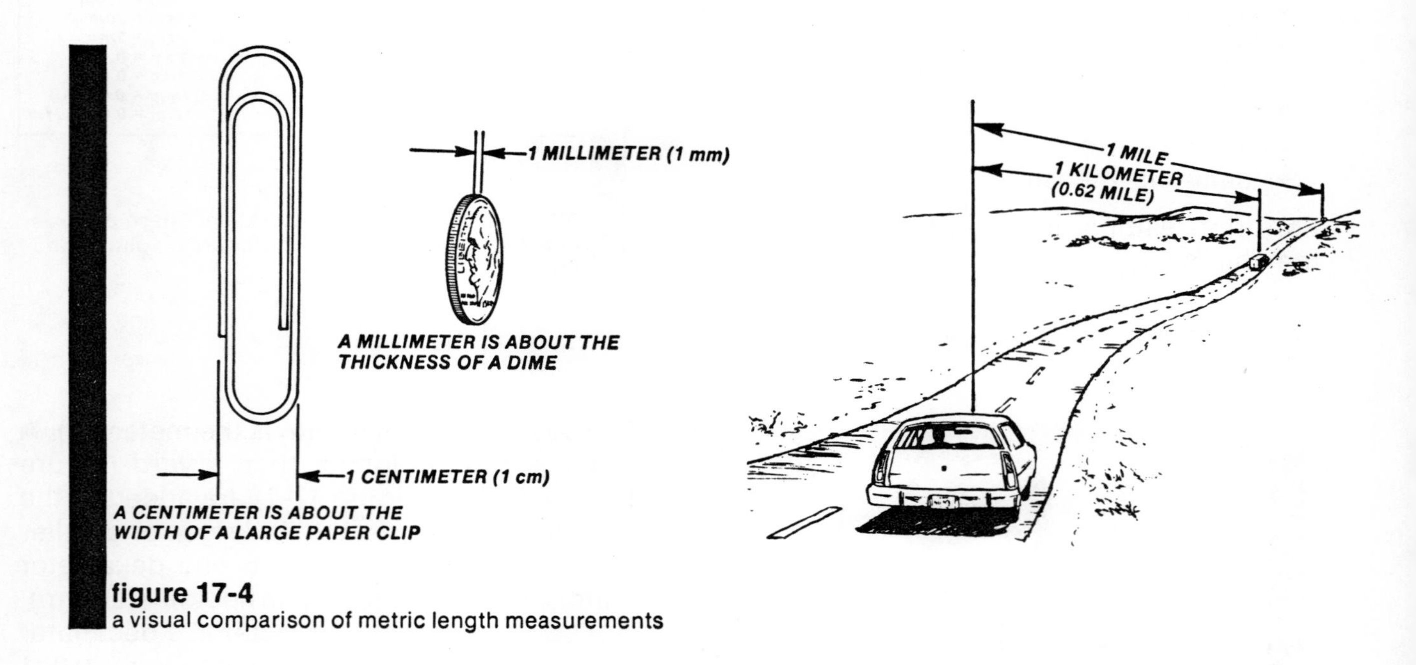

figure 17-4
a visual comparison of metric length measurements

meters if possible. Machining drawings normally have dimensions in *millimeters*.

Most metric drawings made in the United States are labeled METRIC. You can easily identify a metric drawing by the size of the dimensions. Millimeter dimensions are about 25 times larger than inch dimensions. Figure 17-4 will help you to remember the basic sizes.

Figure 17-4 gives you a chance to relate basic metric units to everyday objects.

self-check quiz 17-a

1. Measure the following bars. Express answers in cm and mm.
   a.
   b.
   c.
   d.

2. Draw lines the required length.
   a. 1 mm
   b. 10 mm
   c. 1 cm
   d. 2.5 cm
   e. 1 in.
   f. 1/4 in.
   g. 5 cm
   h. 50 mm
   i. 0.5 cm
   j. 0.01 m
   k. 0.05 m

3. Measure the following objects:
   a. Length and width of a regular size paper clip ________ mm
   b. Wire diameter of a regular size paper clip ________ mm
   c. Length and width of a large paper clip ________ cm
   d. Wire diameter of a large paper clip ________ cm
   e. Diameter and thickness of a dime ________ mm
   f. Diameter and thickness of a penny ________ mm

first angle orthographic projection

Thus far we have considered only third angle orthographic projection.

Many countries using metric measurement make their drawings in *first angle orthographic projection.* In first angle projection, views are projected *through* the object onto an imaginary transparent box (Figure 17-5A). The box is then unfolded (Figure 17-5B). Notice that views for first angle projection are the same as for third angle projection, but are located in different positions on the drawing.

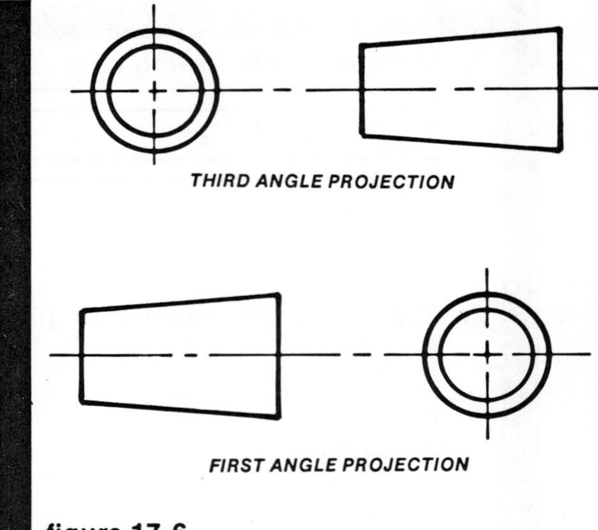

figure 17-6
ISO projection symbols indicating the system of projection used on a drawing

figure 17-5
position of views in first angle projection

So far, metric drawings in the United States are using our familiar third angle orthographic projections. The drawings will look the same but will use millimeter measure.

The International Organization for Standardization (ISO) recommends that a projection symbol be shown on drawings to indicate whether first or third angle projection has been used (Figure 17-6). The symbol is usually shown in or near the title block.

general metric practices

The following general rules have been established:

1. For numbers smaller than one, precede the decimal with a zero, such as 0.85.
2. Do not add extra zeros to the right of the decimal (unless needed for machining).
3. Group numbers greater than 9999 in sets of three figures. Do not use commas. Example: 46 250 or 1 583 421.5
4. Use the symbol ϕ instead of the term DIA for cylindrical shapes.

self-check quiz 17-b

1. What system of projection is used?
2. What unit of measurement is used?
3. What are the dimensions of the slot?
4. What are the overall dimensions of the bracket?
5. Estimate the overall dimensions in inches
6. What is the hole diameter?
7. What is the tolerance on decimal dimensions?
8. Label each of the views.
9. Determine diameters A and B

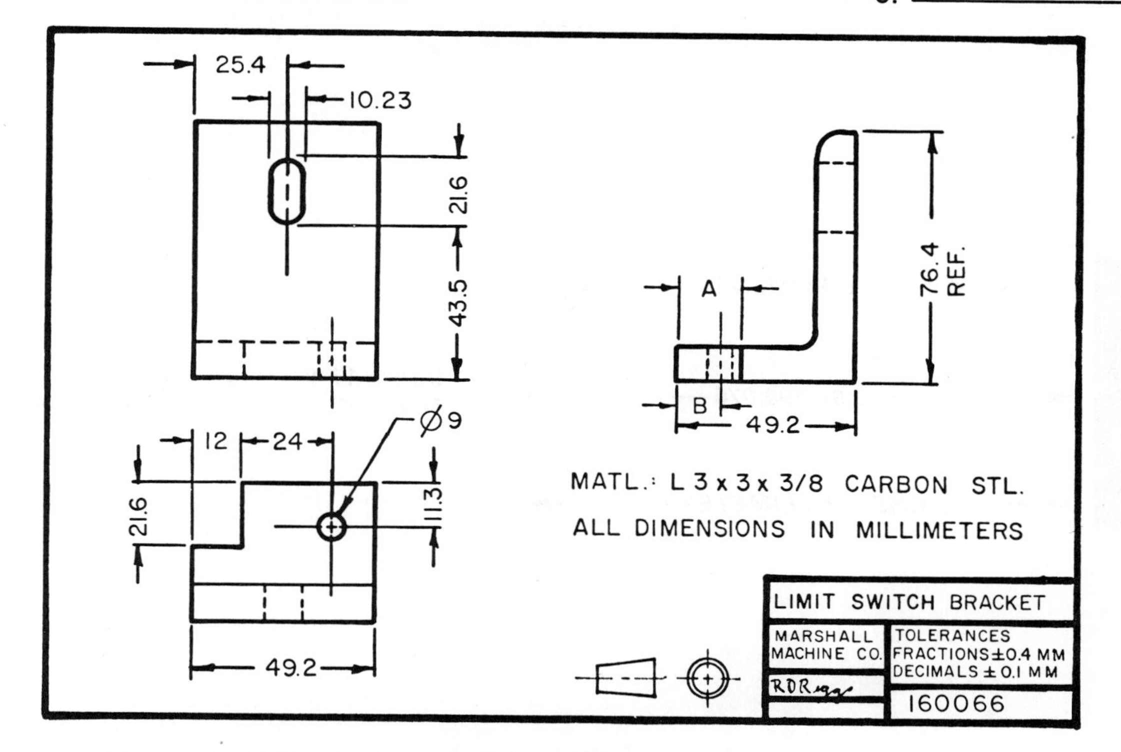

dual dimensioning

During the process of "going metric" in the United States of America, many drawings are being made with dual-dimensions. This means that dimensions are shown in both metric and American Customary units. The dimensions by which a part is designed (known as controlling dimensions) are placed above the dimension lines and the converted values below the dimension lines (Figure 17-7).

Tolerances are shown in a normal manner, adjacent to dimensions (Figure 17-8).

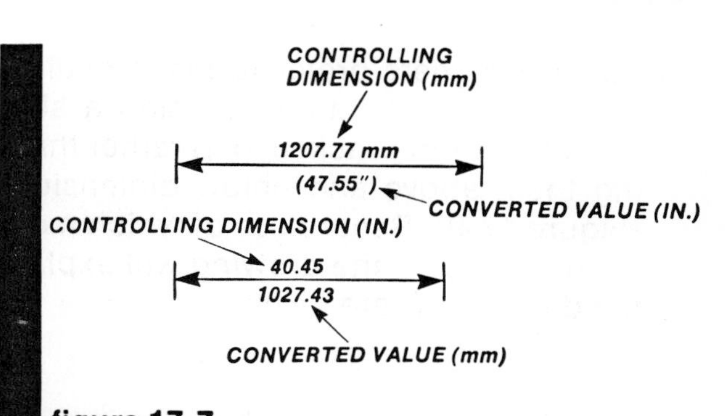

figure 17-7
representing controlling dimensions in dual dimensioning. The unit in which the part is designed is placed above the dimension line. The converted dimension is placed below the dimension line

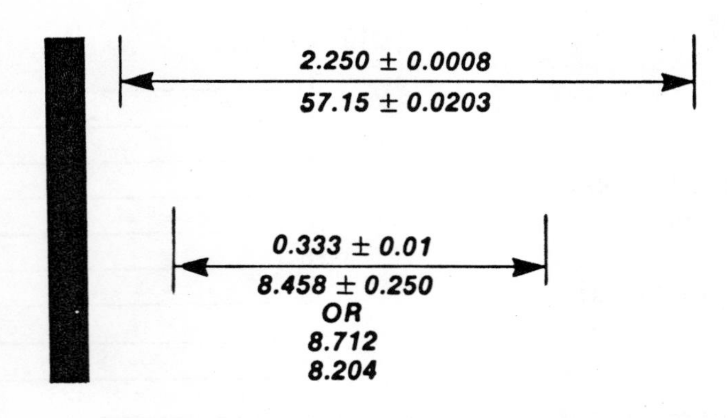

figure 17-8
dual dimensions with tolerances

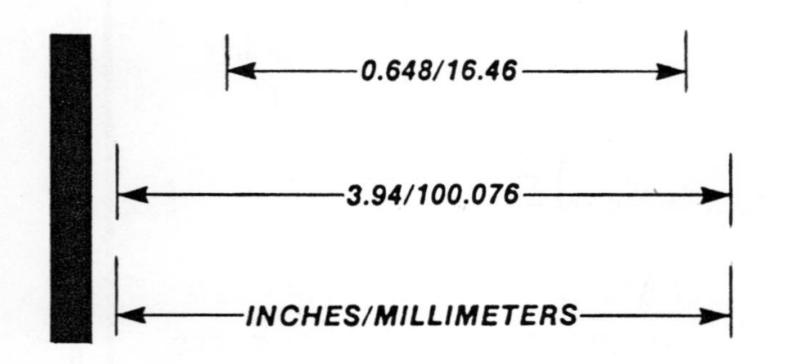

figure 17-9
an alternate method of dual dimensioning

Some designers prefer to place dual dimensions beside one another using a slash (/), parentheses or brackets ([ ]) rather than placing them above and below dimension lines (Figure 17-9). Regardless of which method is used, a note on the drawing will explain how the dimensions are given.

master dimensioning

A complicated drawing can be confusing with dual dimensions. To avoid this problem, the designer may show only the controlling dimensions on the object and include a conversion chart on the drawing (Figure 17-10). This is known as *master dimensioning*. Note that each dimension on the drawing is included in the conversion chart for direct readout in the alternate measuring system.

tabulated dimensioning

A designer may also simplify dual-dimensioning by showing all dimensions on the object with identification letters (Figure 17-11). The dimensions are tabulated as discussed in Unit 10.

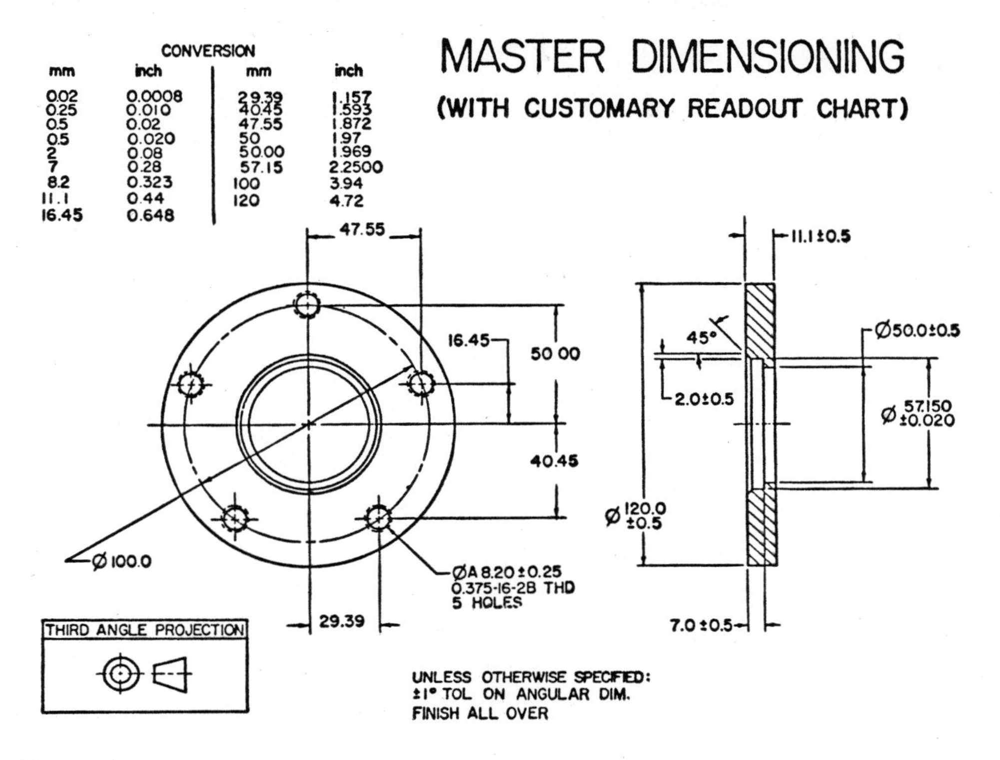

figure 17-10
an example of master dimensioning

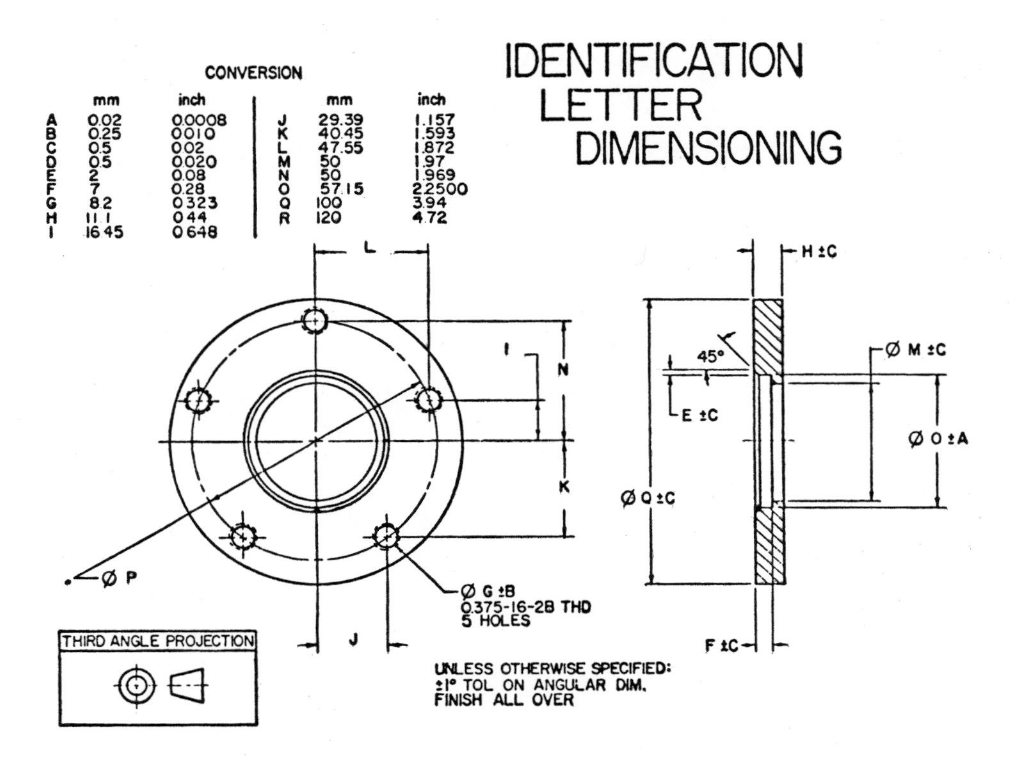

figure 17-11
an example of tabulated (identification letter) dimensioning

This page for student notes

exercise 17-1

______________________ name

1. What system of projection is used for the idler stud?
2. If view A is the front view, What is view B called?
3. Was the idler stud designed in customary or metric units?
4. Determine dimensions C, D, and E in millimeters.
5. Identify surfaces F, G, and H in the other view.
6. If surface W is machined to a dimension of 2.166, must the part be scrapped?

ANSWERS

1. ______
2. ______
3. ______
4. C ______
   D ______
   E ______
5. F ______
   G ______
   H ______
6. ______

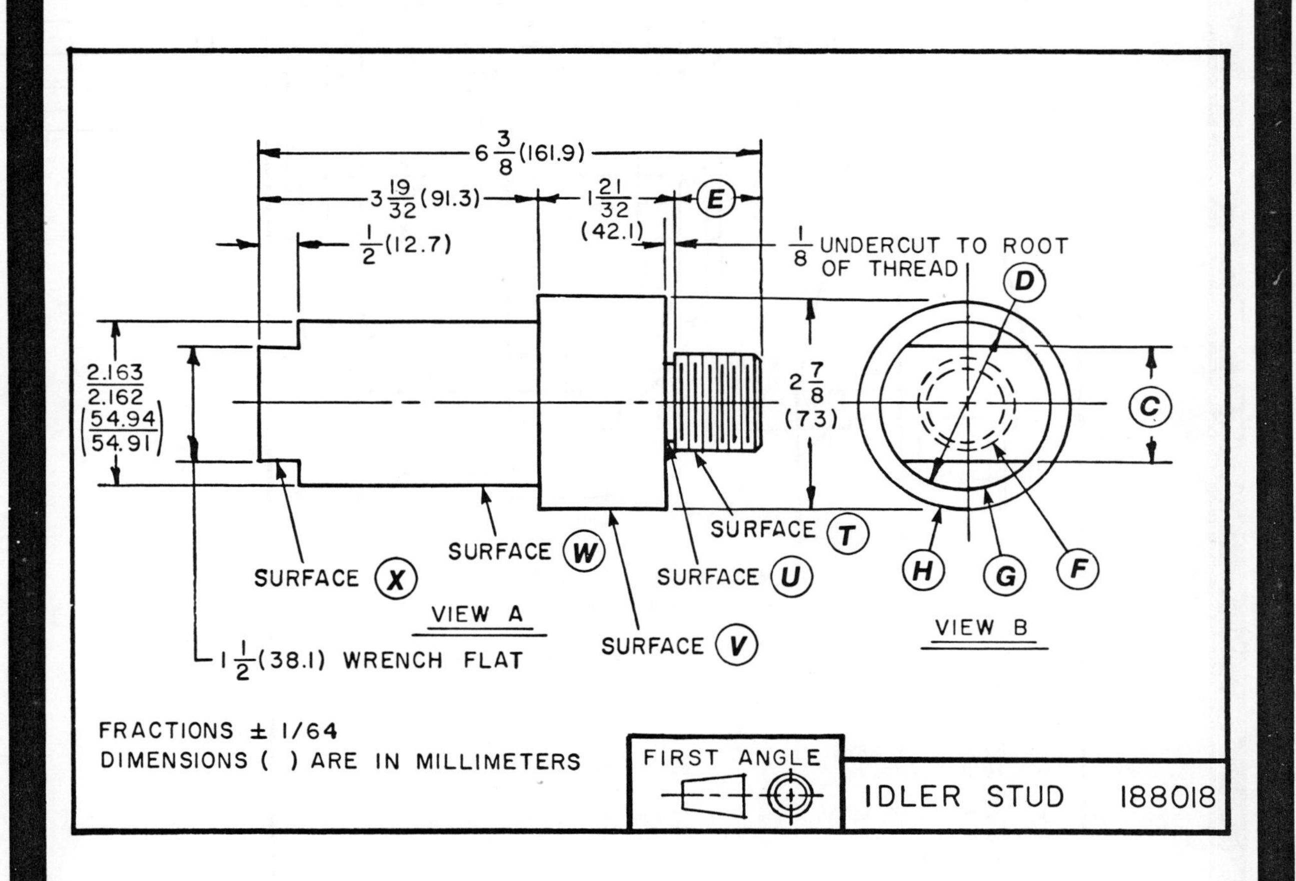

exercise 17-2

________________________ name

1. What system of projection is used?
2. What material is the part made of?
3. What pattern was used to cast the part?
4. What is the tolerance on fractions?
5. Which dimensions are not for machining?
6. Which dimension is expressed by limits?
7. Determine the following dimensions in inches

8. Determine the following dimensions in millimeters

1. ________
2. ________
3. ________
4. ________
5. ________
6. ________
7.
   A ________
   V ________
   H ________
   P ________
   W ________
8. S ________
   T ________
   U ________
   V ________
   W ________

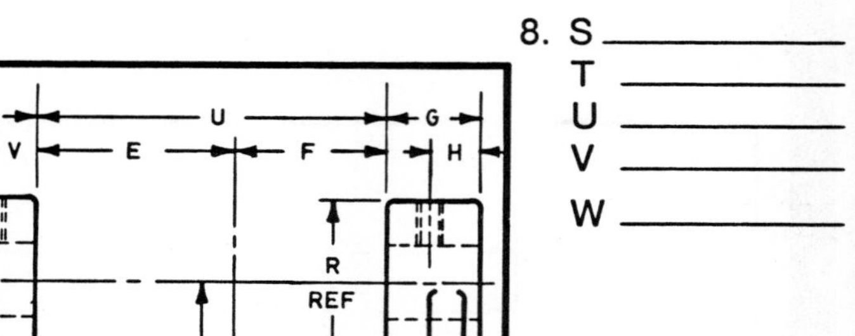

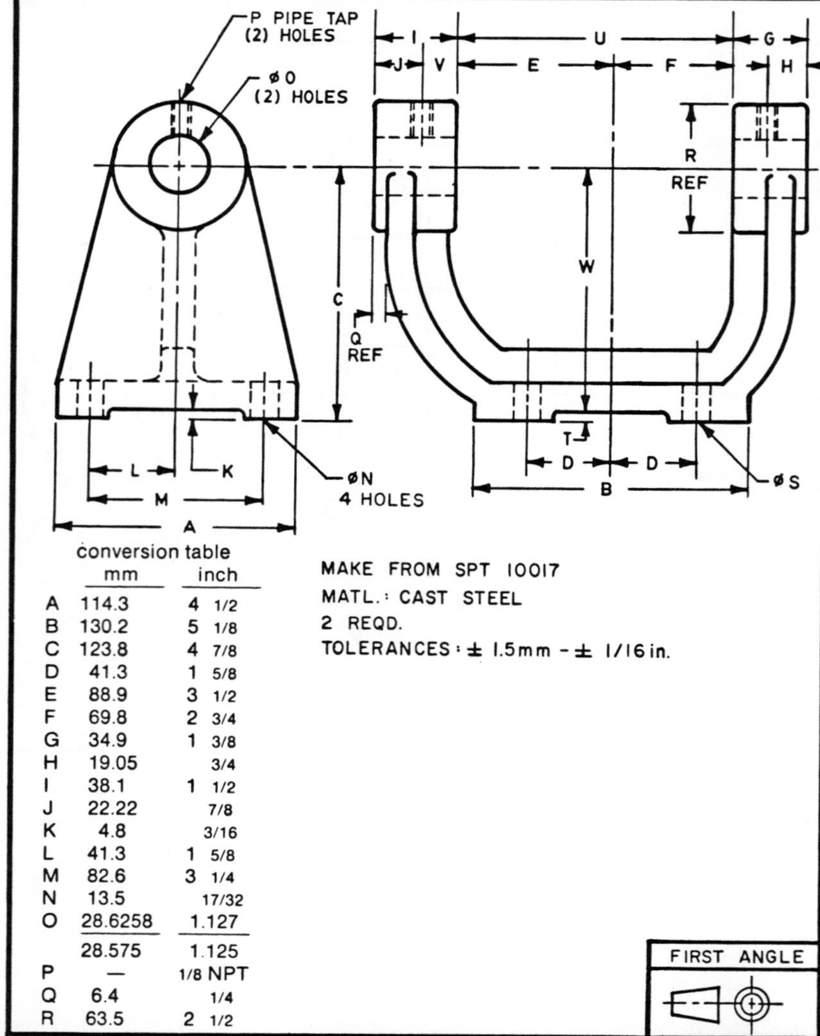

conversion table

	mm	inch
A	114.3	4 1/2
B	130.2	5 1/8
C	123.8	4 7/8
D	41.3	1 5/8
E	88.9	3 1/2
F	69.8	2 3/4
G	34.9	1 3/8
H	19.05	3/4
I	38.1	1 1/2
J	22.22	7/8
K	4.8	3/16
L	41.3	1 5/8
M	82.6	3 1/4
N	13.5	17/32
O	28.6258 28.575	1.127 1.125
P	—	1/8 NPT
Q	6.4	1/4
R	63.5	2 1/2

MAKE FROM SPT 10017
MATL.: CAST STEEL
2 REQD.
TOLERANCES: ± 1.5mm - ± 1/16 in.

exercise 18-1

(Covering Units 1-5)

additional exercises

______________________ name

Sketch several possible front views for each top view.

TOP VIEW

TOP VIEW

TOP VIEW

EXAMPLE

exercise 18-2

(Covering Units 4-7)

______________________________ name

Make a cavalier oblique sketch of the bracket shown in Exercise 13-1 in the space below. Use the corner provided as a starting point. Use a 1:1 scale and maintain good proportion.

exercise 18-3

(Covering Units 4-7)

______________________________ name

Make an isometric sketch of the bracket shown in Exercise 13-1 in the space below. Use the corner provided as a starting point. Use a 1:1 scale and maintain good proportion.

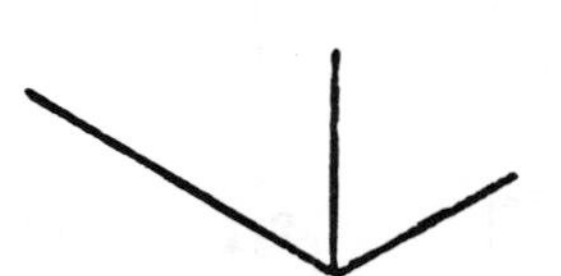

exercise 18-4

(Covering Units 9-12)

______________________________ name

1. Determine dimensions A through I. — 1. A ______ B ______ C ______ D ______ E ______ F ______ G ______ H ______ I ______
2. What is the material of the locator block? — 2. ______
3. Is the right side view a full, half, or partial section? — 3. ______
4. What is the depth of the #10-24 threaded hole? — 4. ______
5. What goes into the 7/8 DIA, 11/8 C'BORE hole? — 5. ______
6. Determine the maximum permissable overall dimensions of the locater block. — 6. ______
7. Which of the following lines is the proper location of the cutting plane for the section view: J, K, L, or M? — 7. ______
8. What is the minimum permissable fit length of the shoulder bushing? — 8. ______
9. If the counterbored hole is located 1.28 inches from line J, is the part acceptable, or must it be scrapped? — 9. ______
10. If the hole for the shoulder bushing measures .844, what must be done to the part? — 10. ______

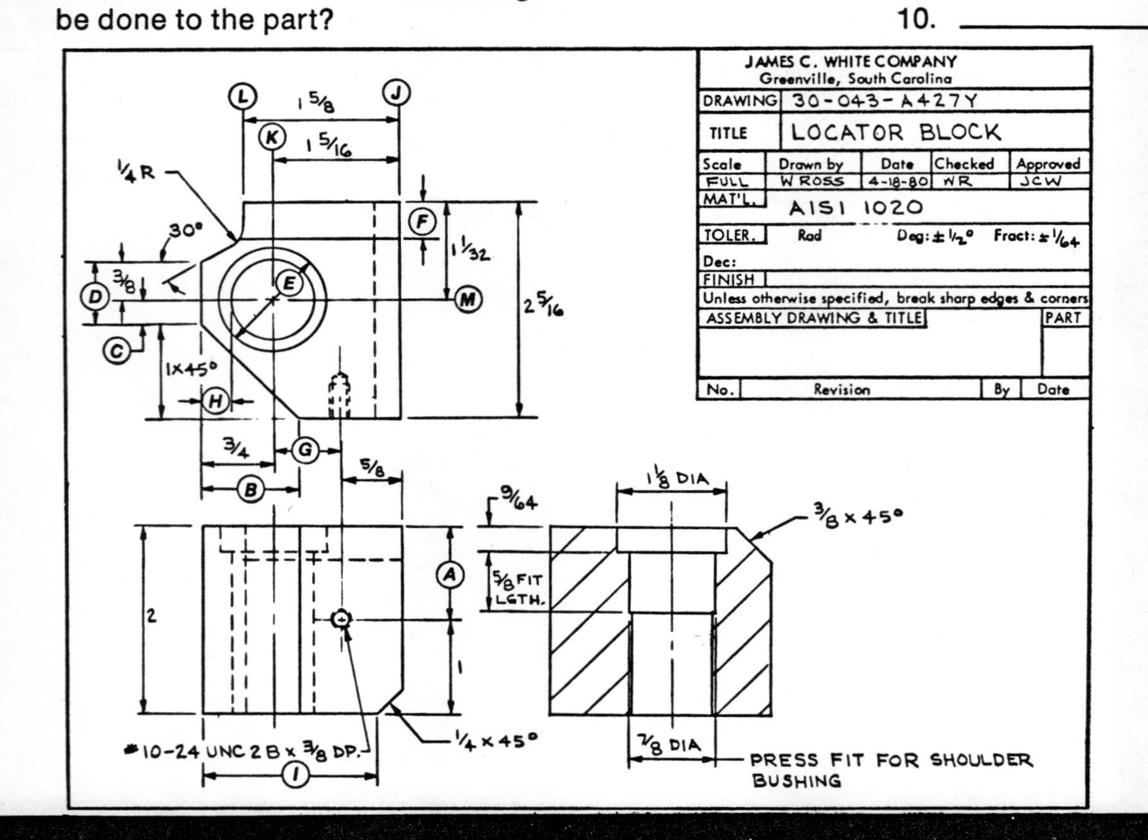

exercise 18-5

(Covering Units 9-12)

______________________ name

1. Determine dimensions A through G.
2. What is the maximum permissible thickness of the part?
3. What is the maximum value of dimension H?
4. What is the minimum value of dimension I?
5. What is the maximum value of dimension J?
6. What is the .375 DIA hole for?
7. How is the .312 DIA hole to be located?
8. What is the machinist to do about sharp corners?
9. What material is the transition to be made from?
10. What is the scale of the drawing?

1. A ______
   B ______
   C ______
   D ______
   E ______
   F ______
   G ______
2. ______
3. H ______
4. I ______
5. J ______
6. ______
7. ______
8. ______
9. ______
10. ______

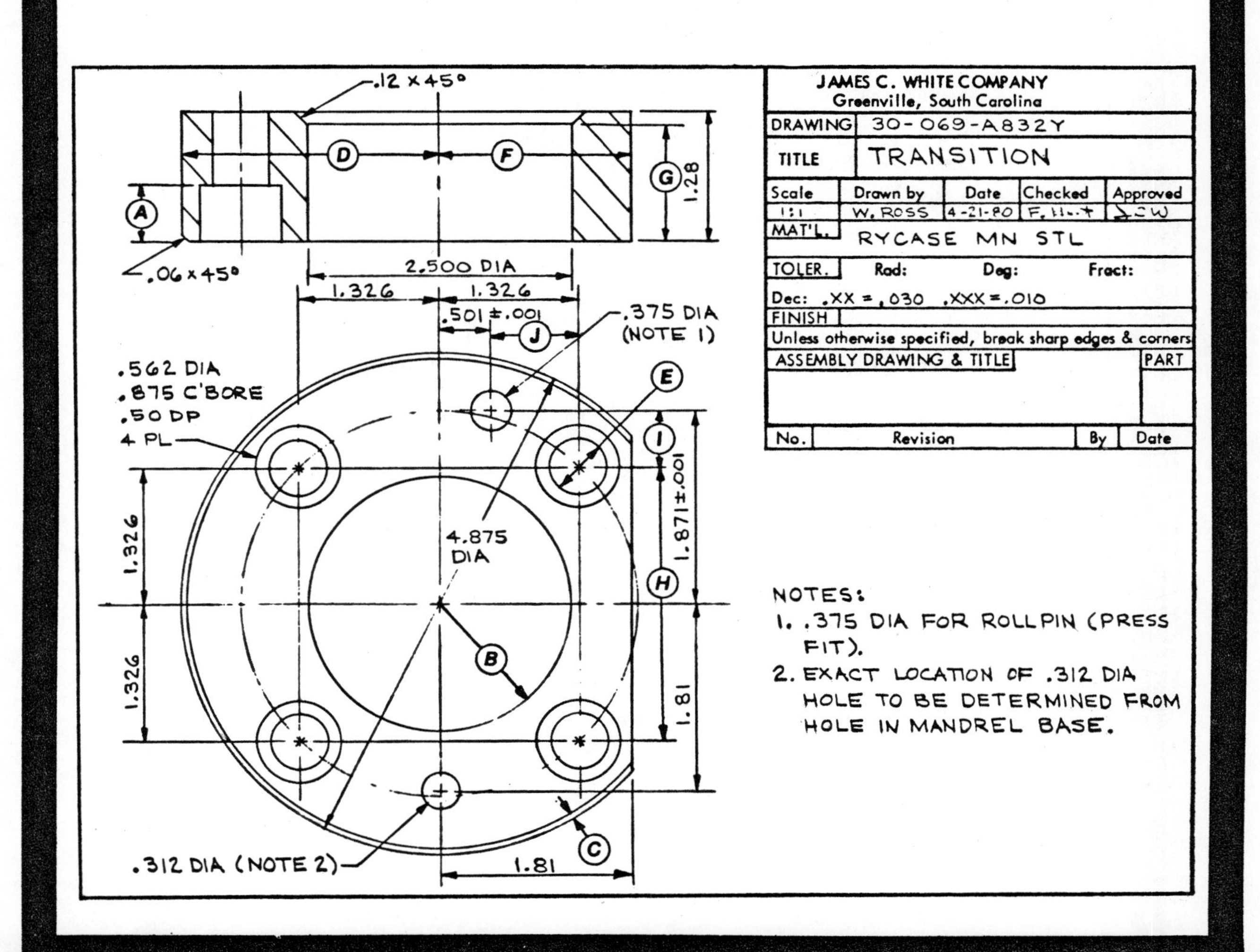

exercise 18-6

(Covering Units 10-16)

_____________________ name

1. What is the material of the JAW?
2. What are the overall dimensions of the JAW?
3. How many holes must be drilled in the JAW?
4. Determine dimensions I through W.

1. ____
2. ____
3. ____
4. I. ____
J. ____
K. ____
L. ____
M. ____
N. ____
O. ____
P. ____
Q. ____
R. ____
S. ____
T. ____
U. ____
V. ____
W. ____

HOLE	HOLE SIZE	LOCATION			COMMENTS
		X	Y	Z	
A1	#10-24UNC-2B,1/2DP	11/32	1 3/8		
A2	↓	1 11/32	1 3/8		
A3	↓	2 9/32	1 3/8		
A4	↓	3 9/32	1 3/8		
B1	3/16 DIA THRU	11/32	2		For Roll Pin
B2	↓	1 11/32	2		↓
B3	↓	13/16	1 3/8		↓
B4	↓	31/32	1 11/16		↓
C1	7/16 DIA, 7/8 DIA C'BORE 5/16 DP	9/16	1 1/16		Toolmaker Verify Location on Vise
C2	↓	3 1/16	1 1/16		
D	1/4 DIA	1 13/16	1 5/16		Drill & Ream for Dowel (Press Fit)
E1	3/8 DIA x 13/16 DP	5/16		5/16	Drill & Ream for Dowel (Press Fit)
E2	↓	3 5/16		5/16	
F1	3/8-16UNC-2B,3/4DP	13/16		11/16	
F2	↓	2 13/16		11/16	
G1	#10-24UNC-2B,5/16DP				Transfer Dimensions From Vise
G2	↓				↓
G3	↓				↓
G4	↓				↓
G5	↓				↓
G6	↓				↓
H1	1/8 DIA x 3/8DP		1 3/8	5/16	Vent Hole for Hole E2
H2	↓		1 3/8	5/16	Vent Hole for Hole E1

Refer back to exercise 18-6

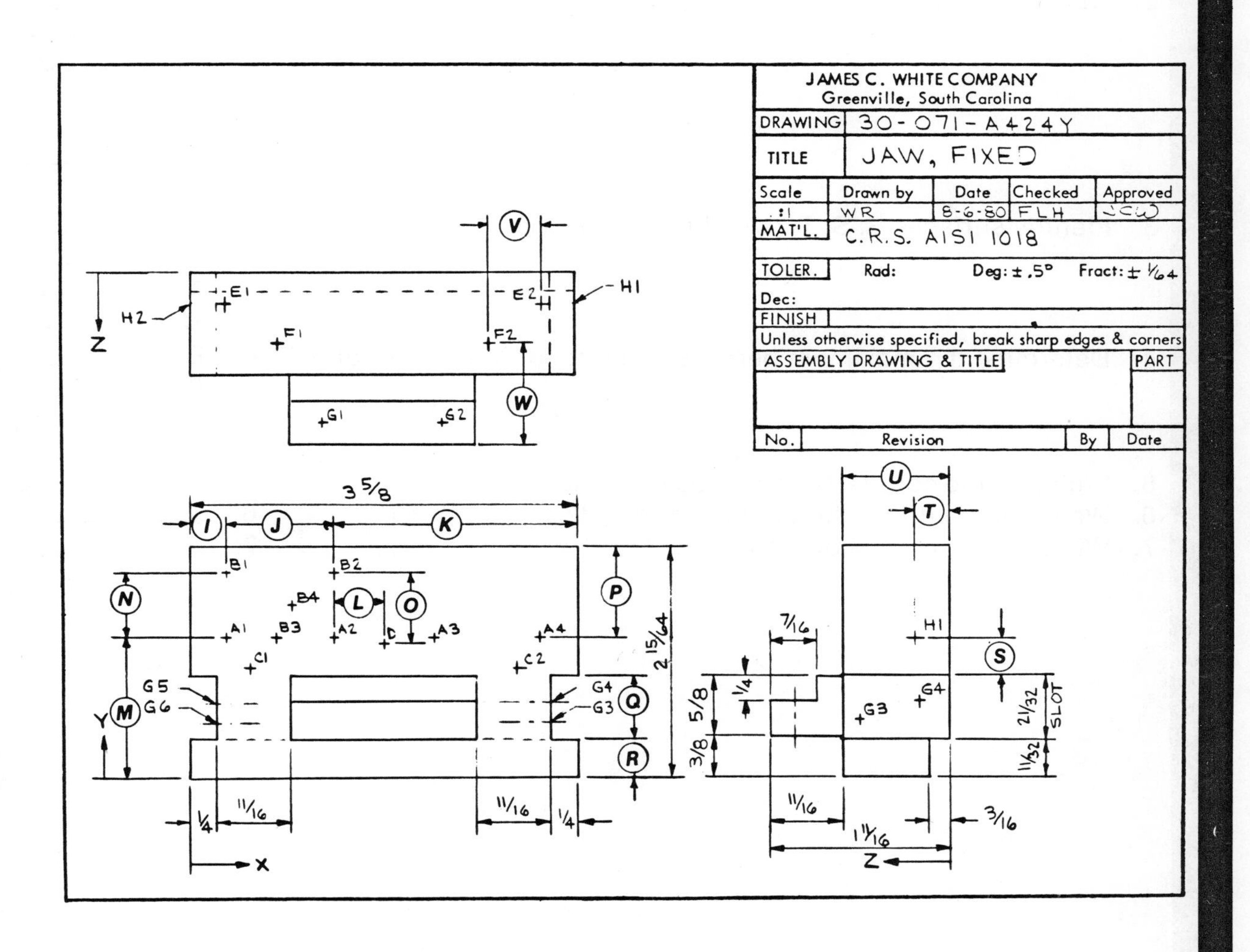

exercise 18-7

(Covering Units 9-15)

______________________ name

1. Determine the overall dimension of the REAR JAW — 1. ________
2. Determine dimensions K through Q — 2. K ________ L ________ M ________ N ________ O ________ P ________ Q ________
3. Identify surfaces R, S, T, and U in another view — 3. R ________ S ________ T ________ U ________
4. Determine the maximum permissable size for holes B, D, and F — 4. B ________ D ________ F ________
5. Which of the counterbored holes is the deepest? — 5. ________
6. Which of the drilled holes is the deepest? — 6. ________
7. What is the purpose of the J holes? — 7. ________

Refer back to exercise 18-7

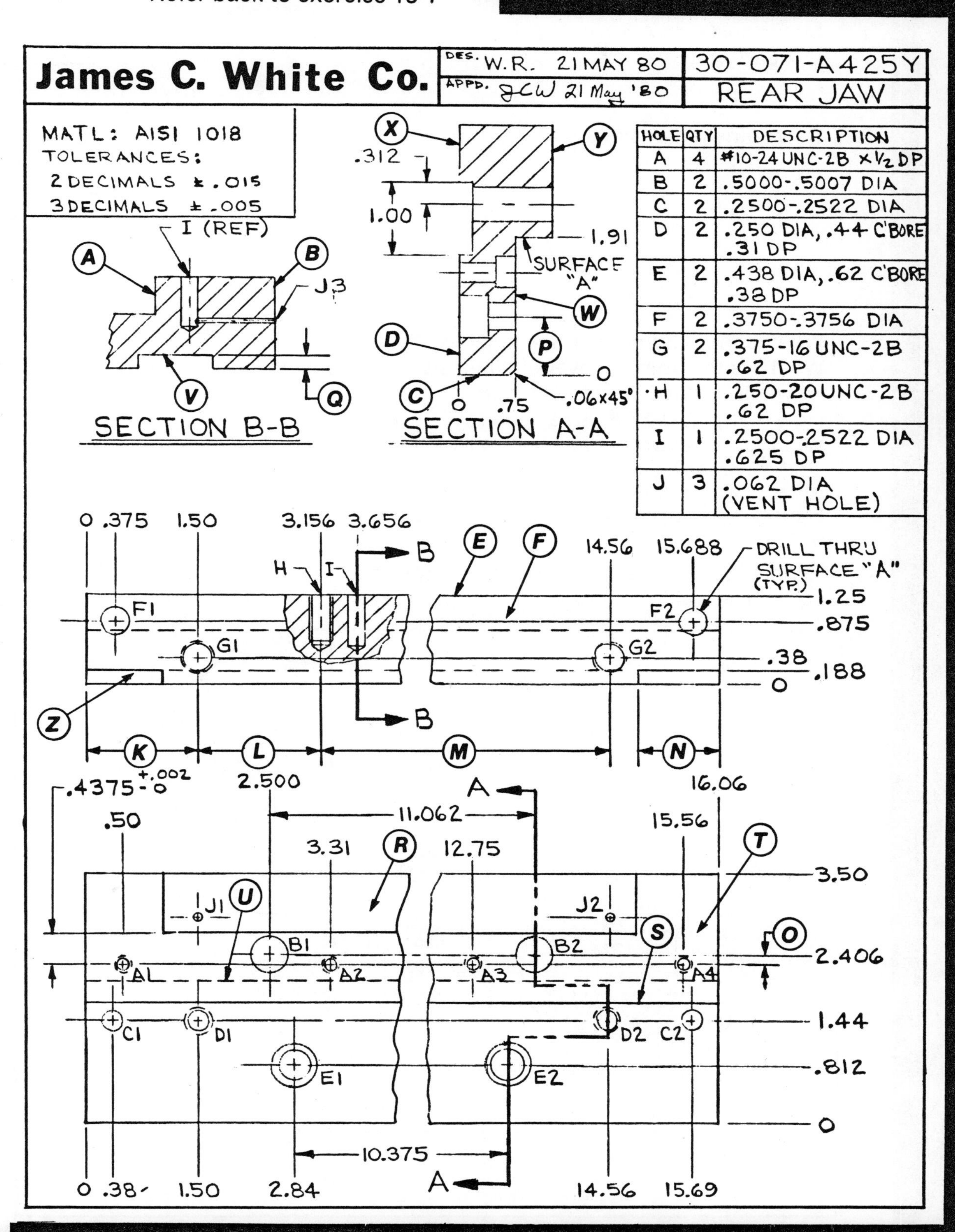

exercise 18-8

(Covering Units 10-15)

__________________________ name

1. Match surfaces A through D in the front view to the top or auxiliary view.
   1. A ________ B ________ C ________ D ________
2. Determine dimensions K through R.
   2. K ________ L ________ M ________ N ________ O ________ P ________ Q ________ R ________
3. What is the required surface finish of the plate? 3. ________
4. What are the nominal overall dimensions of the plate? 4. ________
5. How many #10-32 holes are there? 5. ________
6. How many .250-28 holes are there? 6. ________
7. What is the approximate number of threads in each of the .250-28 holes? 7. ________
8. What class of fit is required for the .250-28 holes? 8. ________
9. Determine the minimum values for S and T. 9. S ________ T ________
10. Express the .3750 DIA. hole size as a nominal dimension with equal bilateral tolerances. 10. ________

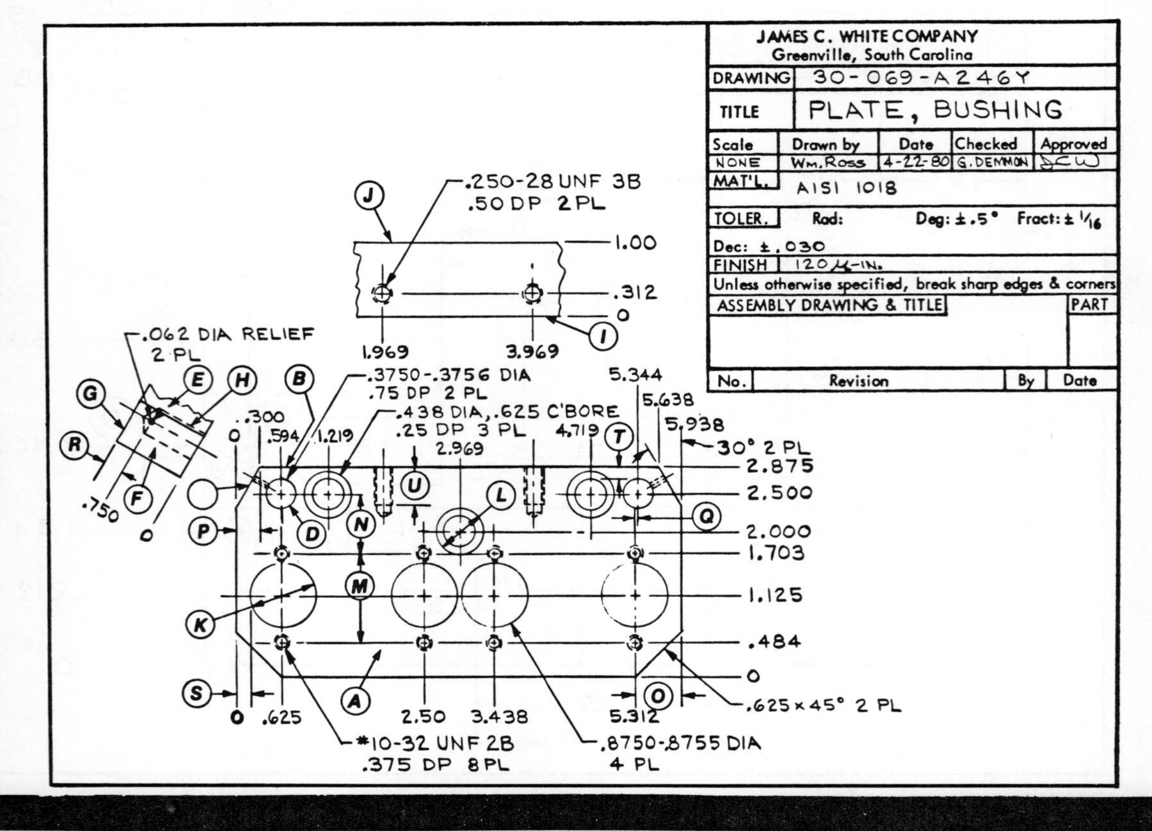

exercise 18-9

(Covering Units 12-16)

_______________________________ name

1. What is the purpose of this drawing? ____________
2. How are the two housings fastened together? ____________
3. List the part numbers of the bearings which support the input shaft. ____________
4. What special machining instruction applies to part 2? ____________
5. What special operation is required for part 4? ____________
6. Which part is longer, part 14 or part 16? ____________
7. Which part is larger, part 6 or part 7? ____________
8. How is the output gear fastened to the output shaft? ____________
9. Which shaft goes faster, part 3 or part 4? ____________
10. What part is used to hold part 6 in part 1? ____________
11. What is the overall depth of the speed reducer, including shafts? ____________
12. What is the center-to-center distance between shafts? ____________
13. What kind of section view is shown? ____________
14. What drawing shows the machining information for the output shaft? ____________
15. What drawing shows the machining information for the input shaft? ____________

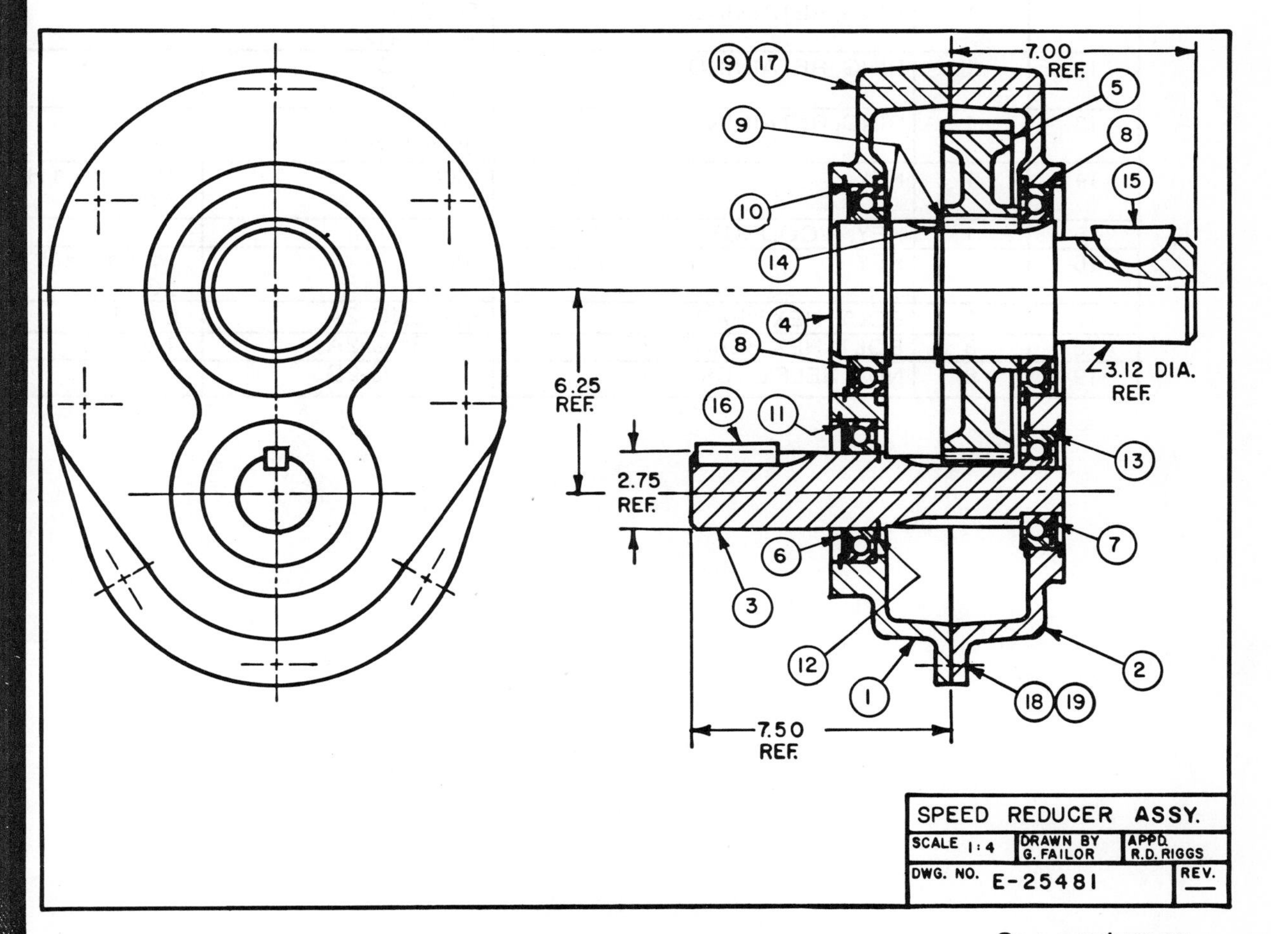

See next page →

← Refer back to exercise 18-9

	BILL OF MATERIAL				
PT. NO.	QUAN.	NAME	MATL.	IDENTIFICATION NO.	NOTES
1	1	HOUSING, INPUT	C.I.	10013-1	
2	1	HOUSING, OUTPUT	C.I.	10013-2	MATCHDRILL W/PT. 1
3	1	SHAFT, INPUT	SAE 1112	10134	
4	1	SHAFT, OUTPUT	A-4140	10027	HEATTREATED BRIN. 225
5	1	GEAR, OUTPUT	SAE 1045	10004-28	
6	1	BEARING		1001-205	
7	1	BEARING		1001-144	
8	2	BEARING		1002-218	
9	2	RING, RETAINING		TRUARC 5100-340	
10	1	RING, RETAINING		TRUARC N5000-354	
11	1	RING, RETAINING		TRUARC N5000-212	
12	1	RING, RETAINING		TRUARC 5100-250	
13	1	RING, RETAINING		TRUARC N5000-175	
14	1	KEY	STL	1004	7/8 W x 5/8 H x 2.00 LG
15	1	KEY, WOODRUFF	STL	1008	
16	1	KEY	STL	1007	1/2 x 1/2 x 2.12 LG
17	5	BOLT, HEX HEAD		1/2-13NC-2A	6.50 LG
18	3	BOLT, HEX HEAD		1/2-13NC-2A	1.50 LG
19	8	NUT, SELF LOCK		1/2-13NC-2B	

trade competency test I

(Covering units 1-4.)

____________________________ name

1. What is the drawing number?
2. What is the name of the object?
3. Name the views shown.
4. Identify the following surfaces or lines in the other view:
5. Name the following lines:
6. What revision is the drawing?
7. Where is the object used?
8. On what drawing are the applied practices located?
9. What was the original height of the block?
10. In the space provided on the drawing neatly sketch top and right side views.

1. ____________________
2. ____________________
3. ____________________
4. A __________________
   B __________________
   C __________________
   E __________________
   F __________________
   G __________________
5. S __________________
   E __________________
   J __________________
   K __________________
   M __________________
   R __________________
6. ____________________
7. ____________________
8. ____________________
9. ____________________

See next page→

← Refer back to trade competency test 1

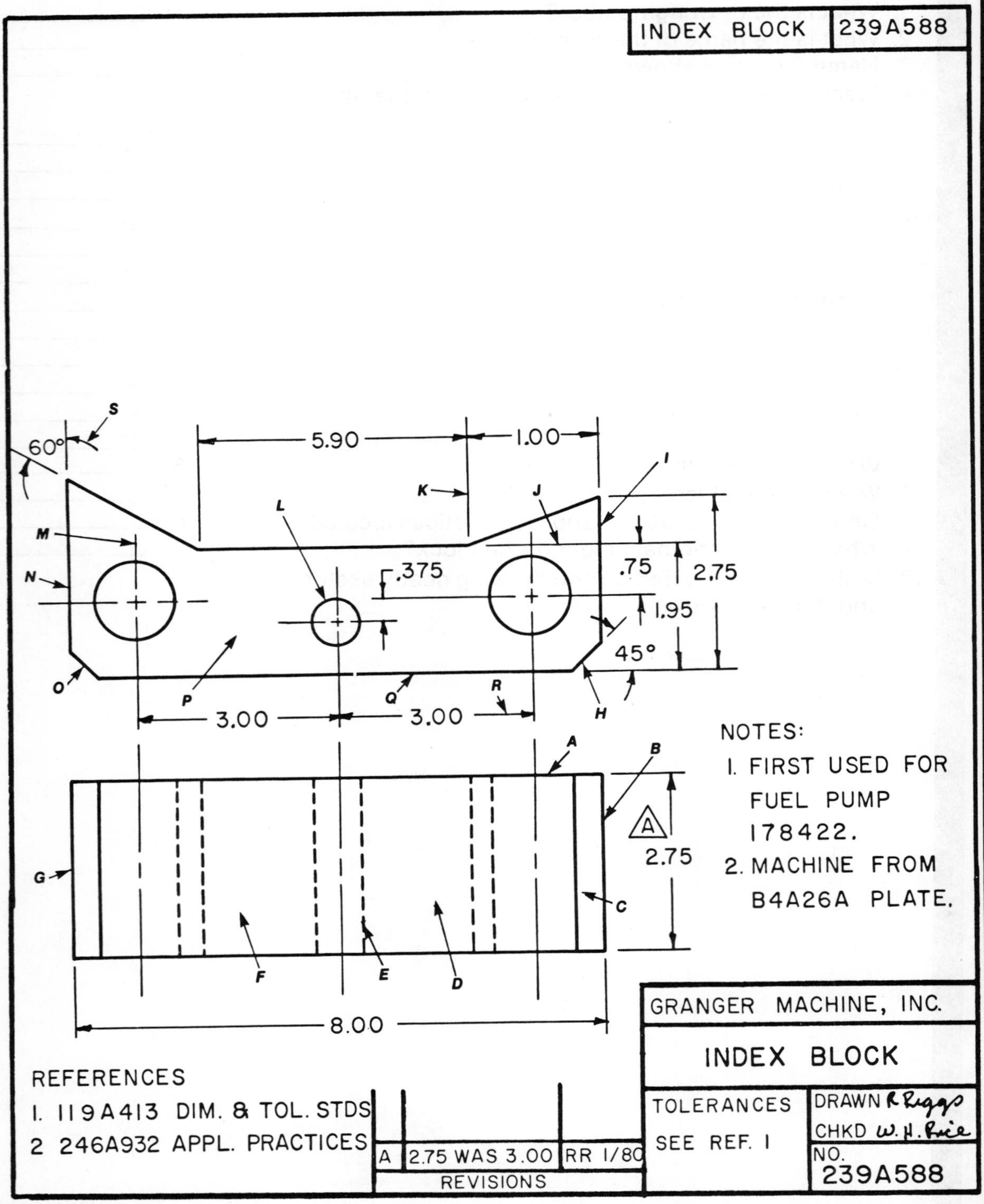

trade competency test II

(Covers Units 5 through 11.)

____________________________ name

1. What is the drawing number? 1. ____________
2. What is the name of the part? 2. ____________
3. What revision is the drawing? 3. ____________
4. What material is the part made from? 4. ____________
5. Name the views shown. 5. ____________
6. List the different kinds of lines used on the drawing. 6. ____________
7. What is the scale of the drawing? What kind of scale is this (architect's, metric,. . . .)? 7. ____________
8. Determine dimensions A through D. 8.
   A ____________
   B ____________
   C ____________
   D ____________
9. Which dimension is not to scale? 9. ____________
10. What is the maximum size of the 4 inch collar? 10. ____________
11. Determine the minimum value for dimension E. 11. ____________
12. If surface F was machined to a diameter of 2.62605, would it be within the allowable tolerance? 12. ____________
13. If surface G was machined to a diameter of 2.5040, would it be within the allowable tolerance? 13. ____________
14. What is the diameter of the bolt circle for the 1/2 - 13 tapped holes? 14. ____________
15. How deep are the 1/2 - 13 tapped holes? 15. ____________

See next page→

← Refer back to trade competency test 2

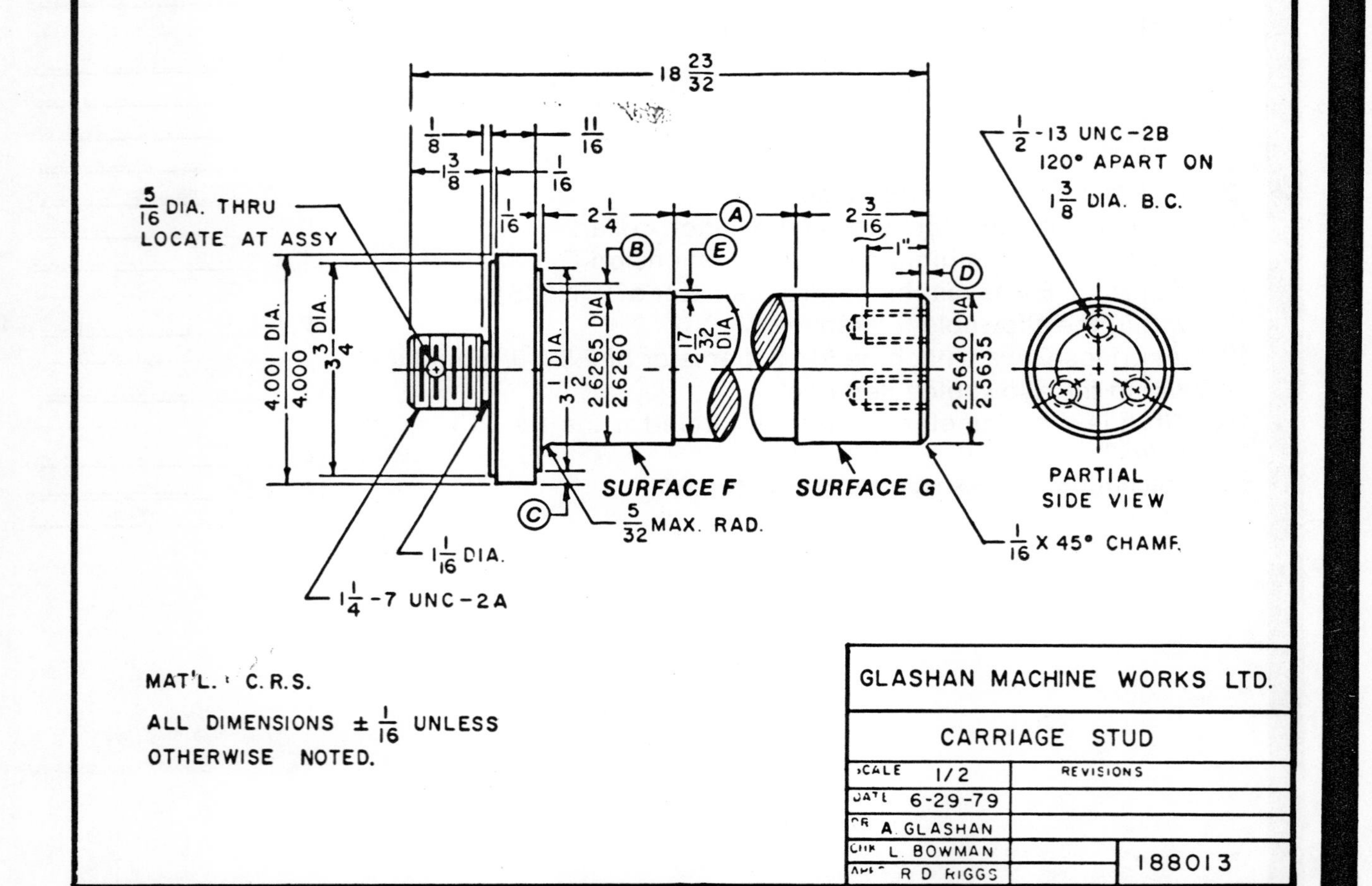

trade competency test III

(Covers Units 12 through 16)

_____________________ name

1. Determine the following:
   A. Drawing title
   B. Drawing number
   C. Drawing revision
   D. Drawing scale
2. Are the dimensions aligned or unidirectional?
3. Which dimension is not to scale?
4. What size are the spotfaced holes?
5. How deep are the spotfaced holes?
6. What goes into the spotfaced holes?
7. How many tapped holes are there?
8. What is the bolt circle diameter for the tapped holes?
9. How deep are the tapped holes?
10. How many full threads are there in one of the tapped holes?
11. What is the upper limit of the bore?
12. What is the nominal diameter of the bore?
13. What is the lower limit of the 4 1/2 diameter?
14. What is the lower limit of the 7 11/16 dimension?
15. Which dimension has a unilateral tolerance?
16. What kind of section is shown?
17. What is the surface finish of the bore?
18. What is the depth of the keyway?
19. What special care must be taken when machining the keyway, spotfaced holes, and tapped holes?
20. What is the small hole (#13 drill) for?

1. A ________
   B ________
   C ________
   D ________
2. ________
3. ________
4. ________
5. ________
6. ________
7. ________
8. ________
9. ________
10. ________
11. ________
12. ________
13. ________
14. ________
15. ________
16. ________
17. ________
18. ________
19. ________
20. ________

See next page

← Refer back to trade competency test 3

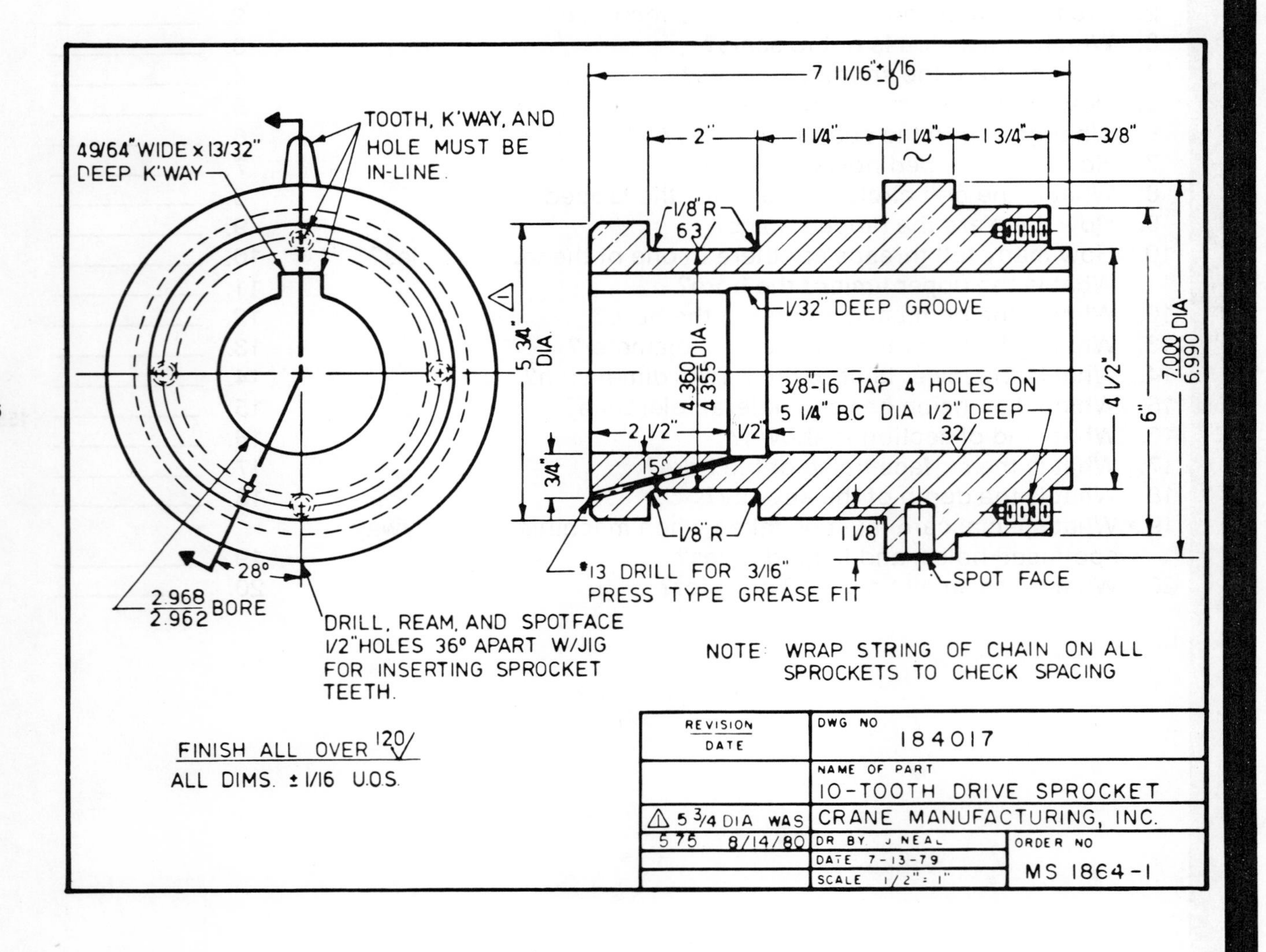

answers to self-check quizzes

self-check quiz 1-a

1. b
2. c
3. d
4. D
5. B
6. D
7. A
8. E
9. F
10. engineers, architecs, designers, drafters, tech. writers, trade workers

self-check quiz 2-a

1. informal
2. formal
3. informal
4. formal
5. formal
6. 237A1042
7. bearing housing
8. A
9. A
10. view 3a added: 1/16 changed to 3/32.

self-check quiz 2-b

1. true
2. true
3. false
4. true
5. true
6. false
7. true

self-check quiz 3-a

A) object
B) center
C) center
D) hidden
E) center
F) object

self-check quiz 3-b

A) center line used as extension line
B) center line
C) object line
D) dimension line
E) center line used as extension line
F) phantom line
G) center line
H) hidden line

self-check quiz 4-a
SEE SEPARATE SHEET

self-check quiz 4-b
SEE SEPARATE SHEET

self-check quiz 5-a

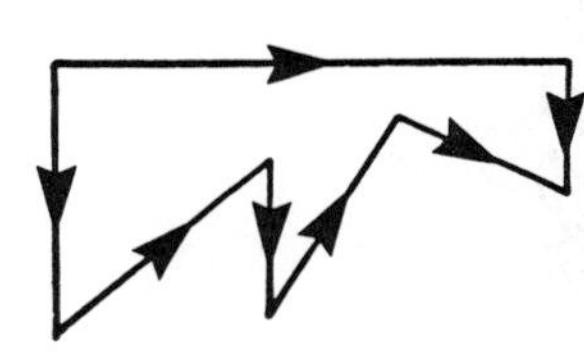

self-check quiz 6-a

1. isometric
2. cabinet oblique
3. cavalier oblique
4. isometric

self-check quiz 7-a

A) 1/4″
B) 5/8″
C) 15/16″
D) 1 1/2″
E) 1 1/16″
F) 2 13/16″
G) 3 5/32″
LM) 1′-0″
KM) 1′3″
IM) 1′8 1/2″
JL) 7″
HN) 2′-11 1/4″
KL) 3″
IK) 5 1/2″
O) 150 mi
P) 270 mi
Q) 330 mi
R) 490 mi

self-check 7-b

1. 8/10, 13/16
2. 1 9/10, 1 14/16 (1 7/8)
3. 2 8/10, 2 13/16
4. 3 7/10, 3 12/16 (3 3/4)
5. 5, 5

self-check quiz 8-a

1. 1859
2. 73,026
3. 301
4. 7484
5. 8244
6. 3,862,604
7. 20
8. 479
9. 1 3/8
10. 1 11/16
11. 25 13/16
12. 3/4
13. 3 3/8
14. 15/32
15. 2 47/64
16. 3 3/8
17. 100
18. 1/100
19. 50
20. 22.15
21. 483.997
22. 141.61
23. 1767.744
24. 1751.4
25. 16.425
26. 16440.32
27. 9.69
28. 26.98
29. 44.206
30. .3678

self-check quiz 8-b

1. 9.325
2. 68.466
3. 758.25
4. 114.25
5. 1.925

self-check quiz 8-c

1. 108
2. 12000
3. 17931
4. .945
5. 7.875
6. 11.375

self-check quiz 8-d

1. 508
2. 1073.28
3. 2.35
4. 7.4
5. 13.5
6. 5.67

self-check quiz 8-e

1. 525.95
2. 7902.55
3. 1 1/6
4. 1 59/80
5. 27/32
6. .001
7. .1
8. .0001
9. .01
10. .2
11. .05

12. .0003
13. .005
14. .6
15. .0012

self-check quiz 9-a

1. 81/2 × 5 × 23/4
2. 2
3. 21/2
4. 5/16
5. 21/2
6. 11/2
7. 5
8. 11/4
9. 29/16
10. 21/2, 5/16, 31/2, 1/4, 11/2, 81/2, 2.50, 1/2, 7/16, 23/4, 5

self-check quiz 9-b

1. unidirectional
2. A = 1″
   B = 1/4
   C = 1/8
   D = 45°
   E = 1/2
3. No
4. 11/4 DIA × 21/4 high
5.

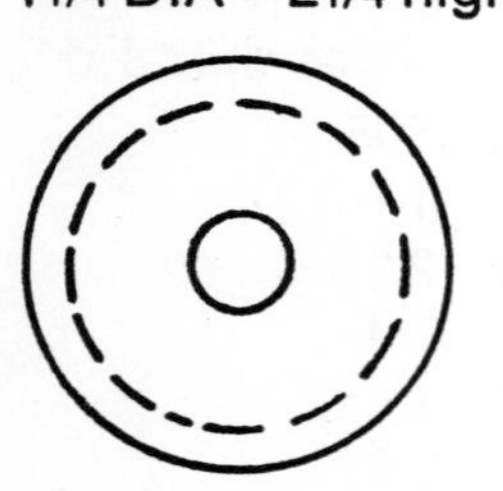

6. same as front view

self-check quiz 10-a

1. 1/2
2. 5/8
3. 5/16
4. 90°
5. 21/30
6. A = 1/2
   B = 21/32
   C = 5/8
   D = 1/4
   E = 5/8

self-check quiz 10-b

1. E and B

self-check quiz 11-a

1. 30°±1/2°, 1/2±1/16, 1.428±.010, .714±.010
2. 2.50
3. 1/2 21/2
4. .228-.240, .500/.508
5. .228-.240 DIA, 21/2 , 2.50 ,30°±1/2°

self-check quiz 12-a

1. A
2.

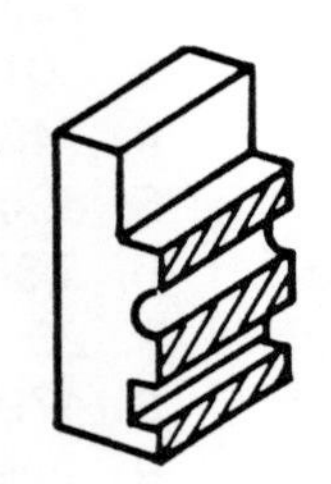

self-check quiz 12-b

A = 1/2, B = 1″, C = 11/32, D = 13/8, E = 9/32, F = 23/4, G = 13/4, H = 13/8

self-check quiz 13-a

1. D
2. C
3. B
4. A

self-check quiz 14-a

1. simplified
2. unified
3. 3
4. external
5. 28
6. 14
7. fine
8. 1/4

self-check quiz 17-a

1. 5.5cm, 55mm
   2.7cm, 27mm
   10.3cm, 103mm
   1.3cm, 13mm
2. SEE SEPARATE SHEET

3a. 32 × 12
b. 0.8
c. 4.8 × 0.95
d. 0.1
e. 17.5 × 1
f. 19 × 1.3

self-check quiz 17-b

1. first angle
2. mm
3. 10.23 × 21.6
4. 49.2 × 49.2 × 76.4
5. 2 × 2 × 3
6. 9
7. j 0.1mm
8. (SEE SEPARATE SHEET)
9. A = 21.6
   B = 11.3

self-check quiz 4-a

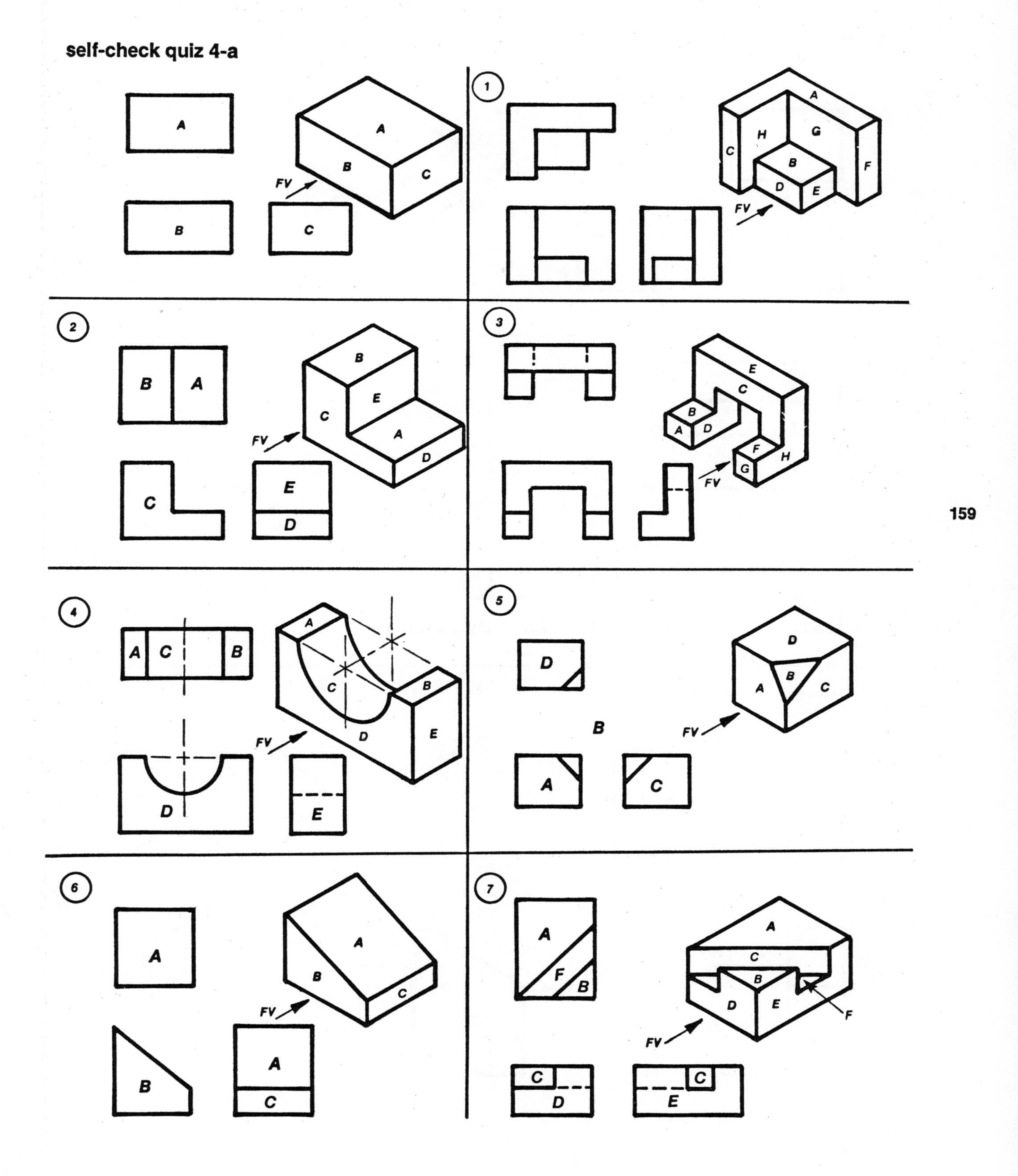

self-check quiz 4-b

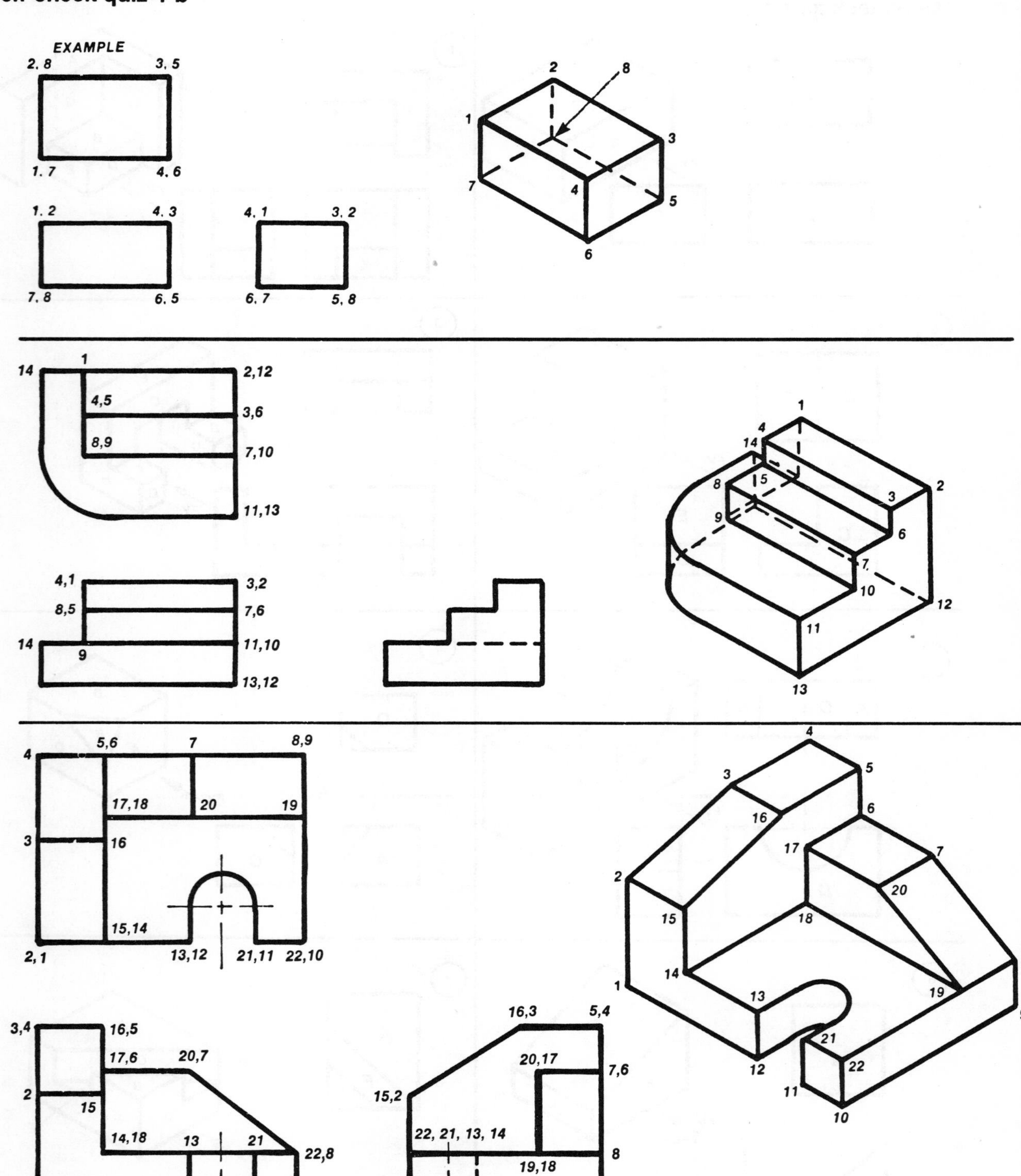

self-check quiz 17-a

2. a. _

   b. _____

   c. _____

   d. _____________

   e. _____________

   f. ___

   g. __________________________

   h. __________________________

   i. __

   j. _____

   k. __________________________

self-check quiz 17-b

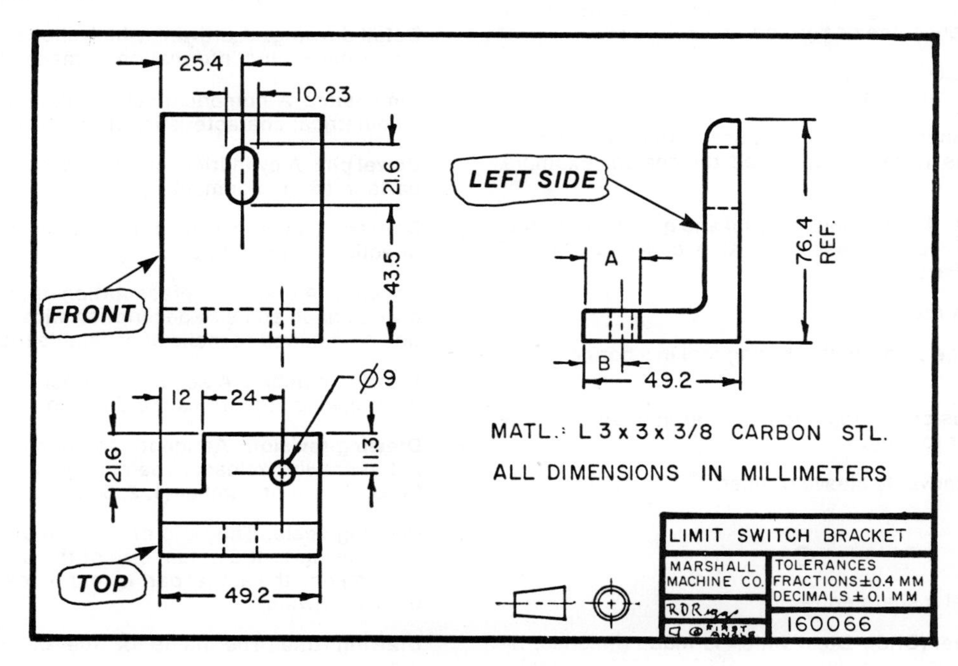

GLOSSARY

A

Architect's scale: A rule used primarily for drawing building plans and details. Sometimes used for machine drawings. Fractions of an inch on an architect's scale represent feet on the actual object.

Assembly drawing: A drawing showing a number of parts or subassemblies joined together.

Auxiliary view: A view of a surface not in one of the normal planes of projection.

B

Bar stock: Metal bars in standard shapes (round, square, hexagonal, octagonal).

Baseline dimensioning: Measuring all dimensions from a common reference line or surface.

Basic size: That dimension to which tolerances are applied to determine limits of size.

Bilateral tolerance: A tolerance which permits variation in both directions from the specified dimension. Example: 1.380 ± .020

Bill of material: A list giving material and quantity information for an object or assembly. Also called a "List of Material" or "Parts List".

Blend: See *Feather.*

Bore: To enlarge a hole with a boring tool in a boring mill or lathe.

Boss: A protrusion or projection circular in shape, usually on a casting or forging.

Bushing: A removable sleeve or liner.

C

Chamfer: Bevel a sharp edge.

Cold rolled steel (CRS): Steel which is made (finished) by cold rolling. Usually low carbon steel (.12 - .20% carbon).

Counterbore: Enlarge a hole for part of its depth. The counterbored portion serves as a recess for a bolt head or nut.

Countersink: Machine a depression to fit the conical head of a screw.

Cutting plane: An imaginary plane which passes through an object to show an internal section.

D

Datum: A reference point, line, or plane from which locations of features on an object are established.

Datum dimensioning: See *Baseline dimensioning.*

Designer: A person with considerable drafting experience who creates design drawings under the direction of an architect or engineer.

Detail drawing: A drawing which shows all the manufacturing or machining information to make an object.

Dimension: A numerical value defining the size or shape (geometrical characteristic) of an object.

Dowel pin: A cylindrical or tapered rod used to hold mating parts in fixed alignment.

Drafter: A person who draws detail drawings under the direction of an architect, engineer, or designer.

Drawing: A graphic representation of an object, or objects, giving shape and size description with such other information as required to make or assemble the object or objects.

Drawing number: A sequence of numbers (and sometimes letters) which identifies the drawing.

Drawing revision: An identified change to a drawing after initial drawing release. The change is designated by letter (A, B, C, . . .) or number (1, 2, 3, . . .).

Drawing scale: The ratio of the size of the part shown on the drawing to the actual size of the part. A drawing scale of ¼ means the actual part is four times larger than shown on the drawing.

Drawing title: The name of the object shown on the drawing.

Drill: To make a cylindrical hole in a part with a drill; a pointed cutting tool.

Dual dimensioning: Dimensioning a drawing in inch values with metric equivalents.

E-F

Exploded view: A pictorial view of an assembly which shows how the individual parts go together.

Face: Machine a flat surface perpendicular to the axis of rotation.

Fastener: A device which joins two or more parts together. Nuts, bolts, screws, rivets, and pins are common types of fasteners.

Feather: Round off the edges of the part being machined.

Fillet: A concave radius at the intersection of two surfaces.

First angle orthographic projection: A system of drawing in which surfaces are projected *through* the object onto the picture planes. The top view goes below the front view, and the right side view goes on the left of the front view. Used in Europe.

I-K

Isometric view: A pictorial view which has the horizontal surfaces of an object drawn at 30° to true horizontal.

Key: A piece of steel which fastens a pulley, gear, or coupling to a shaft.

Knurl: Grooves cut in a diamond pattern to provide a gripping surface.

L

Lay: The predominant surface pattern due to machining.

Layout: A preliminary drawing which contains all the basic design information. A formal drawing and details are drawn based on the information on the layout.

Limit dimensions: The upper and lower limits of a dimension expressed directly on a drawing.

Limits: The largest and smallest permissible size of a dimension.

Location dimensions: Those dimensions which define the position of shapes of an object relative to each other.

M-N

Mechanical engineer's scale: A rule graduated for full size, half size, quarter size and eighth size. Fractions of an inch on a mechanical engineer's scale represent inches on the actual object. Often used for machine drawings.

Neck: Cut a groove around a cylindrical object, usually at a change in diameter.

Nomenclature: The name of an object or parts.

Nominal size: A general size used for designation or identification. The actual size may be different from the nominal size. For example, a .506 inch diameter hole could be called a ½ inch (nominal) hole.

O-P

Oblique view: A pictorial view which shows the front surface true size and shape while the top and side surfaces recede at an angle to the horizontal.

Parallel: Equidistant at all points from a datum plane.

Parts list: See *Bill of materials.*

Perpendicular: At 90° (right angles) from a datum plane.

R

Ream: Finish a drilled (or punched) hole very accurately. A rotating fluted tool is used.

Reference: A drawing, specification, or other document invoked by a drawing as part of the drawing requirements.

Reference dimension: A dimension given for information only. Not intended for machining unless it is given elsewhere without the REF designation.

Revision: See *Drawing revision.*

Round: A convex radius at the intersection of two surfaces.

S

Section view: A cross-cut through a part or assembly which shows internal details.

Serrate: To cut grooves close together to create a gripping surface.

Size dimensions: Dimensions which define the physical limits of shapes of an object.

Specification: A document of printed information giving details on fabrication, inspection, material requirements, or other information related to the machining drawing.

Spotface: A circular, flat surface on a casting or forging which provides a flat seat for a nut or bolt.

Surface finish: A measure of surface irregularities on a machined surface created by the machine tool. Surface finish includes roughness, waviness, and lay pattern.

T

Tabular dimensioning: A method of dimensioning in which dimension values are listed in a table on the drawing. Often used when a large number of holes are required on a part. Also used when more than one part of similar shape but different size is required.

Tolerance: The allowable variation from a specified dimension.

Turn: Machine a diameter on a lathe.

U-V

Unilateral tolerance: A tolerance where variation is allowed in only one direction from the specified dimension. Example: $2.250 \begin{matrix} +.005 \\ -.000 \end{matrix}$

Visualization: Forming a mental picture of an object by viewing a two-dimensional drawing.

INDEX

Addition, review of, 60
Adjacent parts, sectioning, 98
Angular measurements, 65-66
Angular tolerances, 89
Architect's scale, 54
Arcs, sketching, 35
Assembly drawings, 123-124
Auxiliary views, 103-107
Baseline dimensioning, 83
Bilateral tolerances, 90
Bill of material, 9
Cabinet drawings, 44
Cavalier projection, 44
Circles, sketching, 35
Common fractions, 63-64
Computer graphics, 125-126
Counterbored holes, 79
Countersunk holes, 80
Cutting plane, 94
Datum, 83
Datum dimensioning, 83
Decimal equivalents, 63-65
Decimal tolerances, 89
Dimensioning
  Curved surfaces, 74
  Holes, 79-83
  Limited spaces, 74
  Out-of-scale, 75
  Tabular, 84
Dimensions, 9, 71-75
Dimensions out-of-scale, 75
Division, review of, 62
Drafting practices, 123-128
Drawing number, 3-4
Drawing revision, 5-6
Drawing scale, 44
Drawing title, 5
Dual dimensioning, 133-134
Engineer's scale, 54
Exploded view, 47
External threads, 109
Fillets, sketching, 37
Finish designation, 117-118
Finish symbols, 118-120
First angle orthographic projection, 132
Fractions, 63-64
Fractional tolerances, 89
Half sections, 95
Inclined surface, 103-104
Internal threads, 109
Irregular shapes, sketching, 36
Isometric views, 43-44
Keys, 113
Keyway, 113
Lay, 118-119
Left-hand thread, 112
Limit dimensions, 74, 90
Limits, 87-90
Lines
  Break, 17
  Center, 13-14
  Cutting plane, 17
  Dimension, 15
  Extension, 15
  Hidden, 13
  Horizontal, 33
  Inclined, 34
  Object, 13
  Leader, 15
  Phantom, 16
  Section, 17, 94
  Vertical, 33-34
Location dimensions, 71
Master dimensioning, 134-135
Mechanical engineer's scale, 54
Metric scale, 55
Metric system, 129-138
Microinch, 117
Multi-part drawings, 123-124
Multiplication, review of, 61
Notes, 7
Number, drawing, 3-4
Numbers, part, 7
Numerical control, 125-126
Oblique views, 44
Offset sections, 96
Ordinate dimensioning, 83
Orthographic projection, 21-31
Part numbers, 7
Parts list, See *Bill of material*
Partial sections, 95
Perspective views, 43
Pictorial views, 43-44
Pictorial sketching, 46
Pins, 113
Pipe threads, 113
Reference dimensions, 72-73
References, 8-9
Removed sections, 97
Revision, See *Drawing revision*
Revolved sections, 97
Ribs, 98
Right-hand thread, 112
Roughness, 117
Rounds, sketching, 37
Rounding off, 65
Section views, 93-98
SI units, 129-130
Size dimensions, 71
Spokes, 98
Spotfaced holes, 80
Steel rule, 53-54
Subtraction, review of, 60
Surface finish, 117-121
Tabular dimensioning, 84, 134-135
Tapped holes, 81
Thin sections, 98
Third angle orthographic projection, 21-23
Thread class, 110
Thread designation, 110, 112
Thread forms, 109, 110
Threaded fasteners, 109-115
Thread series, 109-111
Title block, 5
Tolerances, 87-90
Typical dimensions, 73
Unilateral tolerances, 90
Waviness, 117, 119
Working drawings, 21